HOW THE BODY WORKS

Penguin Random House

Editorial consultant
Dr Sarah Brewer

Contributors
Virginia Smith, Nicola Temple

Project Art Editor
Francis Wong

Senior Editor
Rob Houston

Designers
Paul Drislane, Charlotte Johnson,
Shahid Mahmood

Editors
Wendy Horobin, Andy Szudek,
Miezan van Zyl

Illustrators
Mark Clifton, Phil Gamble,
Mike Garland, Mik Gates,
Alex Lloyd, Mark Walker

Assistant Editor
Francesco Piscitelli

Jacket Editor
Claire Gell

Managing Art Editor
Michael Duffy

Managing Editor
Angeles Gavira Guerrero

Jacket Designer
Natalie Godwin

**Jacket Design
Development Manager**
Sophia MTT

Producer, Pre-production
Nikoleta Parasaki

Producer
Mary Slater

Art Director
Karen Self

Publisher
Liz Wheeler

Publishing Director
Jonathan Metcalf

First published in Great Britain in 2016 by
Dorling Kindersley Limited
80 Strand, London, WC2R 0RL

Copyright © 2016 Dorling Kindersley Limited
A Penguin Random House Company
15 16 17 18 19 10 9 8 7 6 5 4 3 2 1
001–274815–May/2016

A CIP catalogue record for this book

is available from the British Library.

ISBN: 978-0-2411-8801-9

Printed in China

A WORLD OF IDEAS:
SEE ALL THERE IS TO KNOW

www.dk.com

CONTENTS

UNDER THE MICROSCOPE

HOLDING IT TOGETHER

ON THE MOVE

SENSITIVE TYPES

CHEMICAL BALANCE

THE CIRCLE OF LIFE

MIND MATTERS

UNDER THE MICROSCOPE

Who's in charge?

To perform any task, the body's many parts work together in groups of organs and tissues called systems. Each system is in charge of a function, such as breathing or digestion. Most of the time, the brain and spinal cord are the main coordinators, but the body's systems are always communicating and giving each other instructions.

ARE THERE ANY BODY SYSTEMS WE CAN LIVE WITHOUT?

All our body systems are vital. Unlike some organs – such as the appendix – if a whole system fails it usually results in death.

A matter of organization

Systems are communities of body parts with a single function. However, some body parts have more than one job. The pancreas, for example, is part of the digestive system, because it pipes digestive juices into the gut. It also acts as part of the endocrine system, since it releases hormones into the bloodstream.

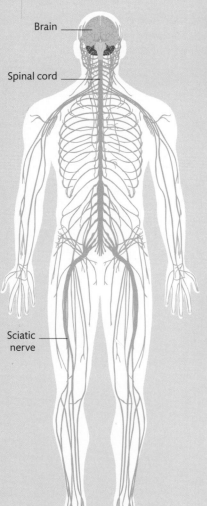

Brain

Spinal cord

Sciatic nerve

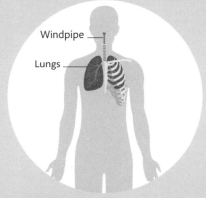

Windpipe

Lungs

Respiratory system
The lungs bring air into contact with blood vessels so that oxygen and carbon dioxide can be exchanged.

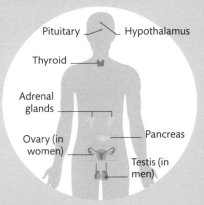

Pituitary

Hypothalamus

Thyroid

Adrenal glands

Ovary (in women)

Pancreas

Testis (in men)

Endocrine system
This system of glands secretes hormones, which are the body's chemical messengers, sending information to other body systems.

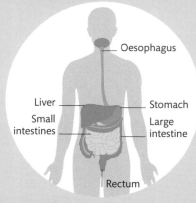

Oesophagus

Liver

Stomach

Small intestines

Large intestine

Rectum

Digestive system
The stomach and intestines are the major parts of this system, which turns food into nutrients needed by the body.

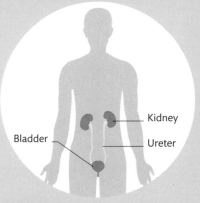

Bladder

Kidney

Ureter

Urinary system
The kidneys filter blood to remove unwanted substances, which are stored in the bladder and expelled as urine.

Central nervous system
The brain and spinal cord process and act upon information received from all over the body through an extensive network of nerves.

Brain
The brain receives information from the eyes, inner ear, and nerves all over the body, which it puts together to get a sense of balance and body position.

Muscles and nerves
Nerve impulses are sent to the muscles to make instantaneous adjustments to body position so as to maintain balance. The nervous system interacts with the muscular system, which in turn acts on the bones of the skeletal system.

Breathing and heart rate
Information from the brain prompts the release of hormones that equip the body for the stress it's undergoing. Breathing becomes more rapid and heart rate increases to carry much-needed oxygen to the muscles.

Digestive and urinary systems
The stress hormones released by the endocrine system act on the digestive and urinary systems to slow them down – energy is needed elsewhere!

78
ONE ESTIMATE OF THE **TOTAL NUMBER OF ORGANS IN THE BODY** – ALTHOUGH OPINIONS VARY!

Everything in balance
None of the body's systems operate independently – they are constantly responding to information they receive to keep the body running smoothly. Each system can make adjustments to compensate for stress placed on other systems, which may require more of the body's resources.

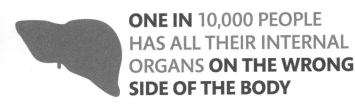

ONE IN 10,000 PEOPLE HAS ALL THEIR INTERNAL ORGANS **ON THE WRONG SIDE OF THE BODY**

OESOPHAGUS

Organs

The organs within the body are typically self-contained and perform a specific function. The tissues that make up that organ help it function in a particular way. The stomach, for example, is largely made of muscle tissue that can expand and contract to accommodate the intake of food.

Stomach structure
Muscle is the main tissue of the stomach, but it is also lined with glandular tissue, which secretes digestive juices, and epithelial tissue, which forms a protective barrier on both the inner and outer surfaces.

Organ to cell

Each organ in the body is distinct and recognizable to the naked eye. Cut through an organ, however, and layers of different tissues are revealed. Within each tissue are different types of cells. They all work together to carry out the functions of the organ.

Stomach has three layers of smooth muscle

STOMACH

Entrance to intestines

Inner wall is lined with cells that secrete mucus or acid

WHICH IS THE LARGEST ORGAN?

The liver is the largest of the internal organs but the skin is actually the biggest organ of the body. It weighs roughly 2.7 kg (6 lb).

Outer layer is covered with epithelial cells

Tissues and cells

Tissues are made up of a group of connected cells. Some tissues come in different types, such as the smooth muscle that forms the walls of the stomach, and skeletal muscle, which is attached to the bones and makes them move. As well as cells, the tissue might contain other structures, such as collagen fibres in connective tissue. A cell is a self-contained living unit – the most basic structure of all living organisms.

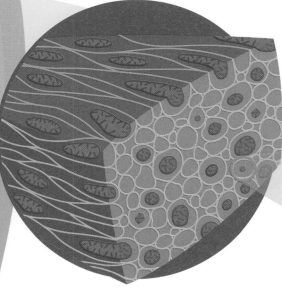

Smooth action
The loose arrangement of the spindle-shaped smooth muscle cells allows this type of muscle tissue to contract in all directions. It is found in the walls of the gut, as well as in blood vessels and the urinary system.

Smooth muscle cells
These long, tapering cells are capable of operating for long periods without tiring.

Tissue types

There are four basic types of tissue found in the human body. These are subdivided into different subtypes, for example, blood and bone are both connective tissues. Each type has different properties – such as strength, flexibility, or movement – that makes it suited to a specific task.

Connective tissue
Connects, supports, binds, and separates other tissues and organs.

Epithelial tissue
Closely packed cells in one or more layers that form barriers.

Muscle tissue
Long, thin cells that relax and contract to create movement.

Nervous tissue
Cells that work together to transmit electrical impulses.

Types of cells

There are around 200 different types of cells in the human body. They look very different under a microscope, but most have common features, such as a nucleus, cell membrane, and organelles.

Red blood cells
Lack a nucleus so they can carry as much oxygen as possible.

Nerve cells
Carry electrical signals between the brain and all parts of the body.

Epithelial cells
Line the surfaces and cavities of the body to form a tight barrier.

Adipose cells
Store molecules of fat that help insulate the body and can be turned into energy.

Skeletal muscle cells
Arranged into fibrous bundles that contract to move bones.

Reproductive cells
The female egg and male sperm combine to form a new embryo.

Photoreceptor cells
Line the back of the eye and respond to light falling on them.

Hair cells
Pick up sound vibrations being transmitted through the fluid of the inner ear.

How cells work

Your body is made up of approximately 10 trillion cells, and each one is a self-contained living unit. Each cell uses energy, multiplies, eliminates waste, and communicates. Cells are the basic units of all living things.

Cell function

Most cells have a nucleus – a structure in their centre that contains genetic data, or DNA. They rely on this data to build various molecules that are essential to life. All of the resources they need to do this are contained within the cell. Structures called organelles carry out specialized functions, similar to the organs of the body. Organelles are held in the cytoplasm, the space between the nucleus and the cell membrane. Molecules are brought into the cell and others are shipped out, just like in an efficient factory.

1 Receiving instructions
Everything that happens in a cell is controlled by instructions in the nucleus. These instructions are exported on long molecules called messenger ribonucleic acid (mRNA) – these molecules travel out of the nucleus and into the cytoplasm.

2 Manufacture
The mRNA travels to an organelle attached to the nucleus called the rough endoplasmic reticulum. There, it attaches to ribosomes that stud the organelle, and the instructions are made into a chain of amino acids that becomes a protein molecule .

3 Packaging
The proteins travel in vesicles – little cellular bubbles – that float through the cytoplasm to the Golgi body. This organelle acts much like the mail room of the cell – packaging the proteins and putting labels on them, which determine where they are sent next.

4 Shipping
The Golgi body places the proteins into different types of vesicles depending on their labelled destination. These vesicles bud off, and those destined for outside the cell fuse with the cell membrane and release the proteins outside of the cell.

Inside a cell
Numerous organelles comprise the internal structure of cells, the types of which vary from cell to cell.

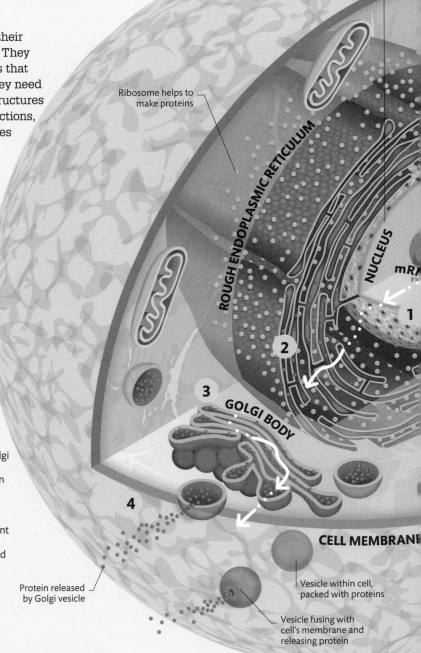

The nucleus is the cell's command centre, containing blueprints in the form of DNA. Surrounding it is an outer membrane, full of pores, which controls what goes in and out.

Ribosome helps to make proteins

ROUGH ENDOPLASMIC RETICULUM

NUCLEUS

mRN

GOLGI BODY

CELL MEMBRANE

Protein released by Golgi vesicle

Vesicle within cell, packed with proteins

Vesicle fusing with cell's membrane and releasing protein

HOW DO CELLS MOVE?

Most cells move by pushing their membrane forward from the inside using long fibres made of protein. Alternatively, sperm cells have tails, which they whip back and forth to move.

Cell death

When cells have reached the natural end of their life cycle they undergo apoptosis – a deliberate series of events that causes the cell to dismantle itself, shrink, and fragment. Cells can also die prematurely as a result of infections or toxins. This causes necrosis, a process in which the cell's internal structure detaches from its membrane, causing the membrane to burst and the cell to die.

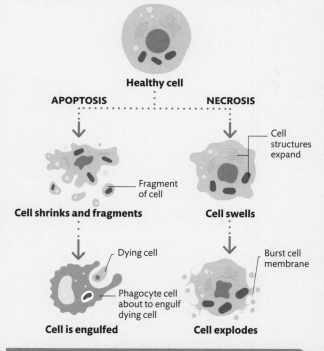

Healthy cell

APOPTOSIS

NECROSIS

Cell structures expand

Fragment of cell

Cell shrinks and fragments

Cell swells

Dying cell

Phagocyte cell about to engulf dying cell

Burst cell membrane

Cell is engulfed

Cell explodes

SMOOTH ENDOPLASMIC RETICULUM

Smooth endoplasmic reticulum produces and processes fats and some hormones. Its surface lacks ribosomes, so it looks smooth.

Centrosomes are the organization points for microtubules – structures that help separate DNA during cell division

Vesicles are containers that transport materials from the cell membrane to the interior and vice versa

Lysosomes act as the cell's clean-up crew. They contain chemicals used to get rid of unwanted molecules.

Cytoplasm – the space between organelles – is filled with microtubules

Mitochondria are the cell's powerhouses, where most of the cell's supply of chemical energy is generated.

VESICLE

MITOCHONDRION

CENTROSOME

LYSOSOME

MOST CELLS HAVE A DIAMETER OF ONLY 0.001 MM

CELL SIGNALLING

A molecule binds to a receptor in a cell's membrane, which triggers events that cause a change in the cell. This signalling molecule may be produced by distant cells, nearby cells, or even the same cell. This is how cells communicate with one another, receive information, and respond to their environment.

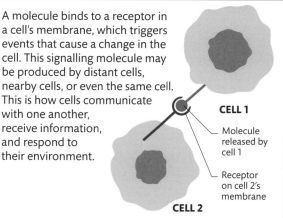

CELL 1

Molecule released by cell 1

Receptor on cell 2's membrane

CELL 2

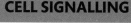

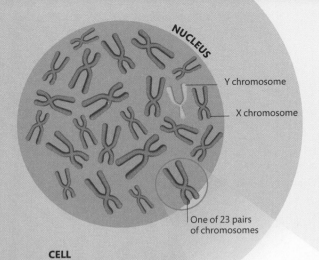

NUCLEUS

Y chromosome

X chromosome

One of 23 pairs
of chromosomes

CELL

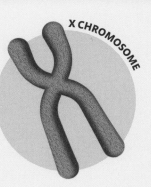

X CHROMOSOME

Y CHROMOSOME

Boy or girl?
Humans inherit one set of 23
chromosomes from their mother and
another from their father. Pairs 1 to
22 are duplicates, but with a slightly
different version of each gene on each
chromosome. Our sex is determined
by our chromosome 23 pairing.
Females have two X chromosomes,
while males have an X and a Y. Few of
the X chromosome genes are repeated
on the shorter Y chromosome, which
mostly carries the genes that produce
masculine characteristics.

Control centre
DNA is stored in the nucleus of
every cell, except for red blood cells,
which lose their DNA as they mature.
In each cell nucleus, there are 2 m (6 ft)
of DNA tightly coiled into 23 pairs of
chromosomes in every cell.

Human library

DNA is a long molecule that provides
all the information necessary for an
organism to develop, survive, and
reproduce. It is like a twisted ladder
with rungs made of a pair of chemical
bases. These bases form long
sequences called genes that are
coded instructions for building
proteins. When a cell needs to
duplicate its DNA or make a new
protein, the two halves of the
ladder unzip so that a copy of the gene can
be made. Humans have more than 3 billion
bases in their DNA and nearly 20,000 genes.

CHROMOSOME

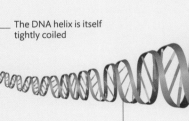

The DNA helix is itself
tightly coiled

Body builders
The genes that build our bodies may range
from a few hundred bases to more than
2 million bases in length – longer than the
small section shown here. Each gene
produces a single protein. These proteins
are the building blocks of the body,
forming cells, tissues, and organs. They
also regulate all the body's processes.

The outer edge of each
strand is made of sugar
and phosphate molecules.

What is DNA?

DNA, also known as deoxyribonucleic acid, is a chain molecule
that exists in nearly all living things. The chain is made up of a
sequence of molecular components, known as bases. Incredibly,
the sequence acts as coded instructions for making an entire
living organism. We inherit our DNA from our parents.

The coloured bars show
the four bases – adenine,
thymine, guanine, and
cytosine – which are
arranged in a particular,
meaningful sequence

Express yourself

The majority of genes are the same in everybody because they code for molecules that are essential for life. However, around 1 per cent have slight variations – known as alleles – that give us our unique physical characteristics. While many of these are harmless traits, such as hair or eye colour, they may also result in more problematic conditions, such as haemophilia or cystic fibrosis. Because alleles come in pairs, one may override the effect of the other so that the trait stays hidden.

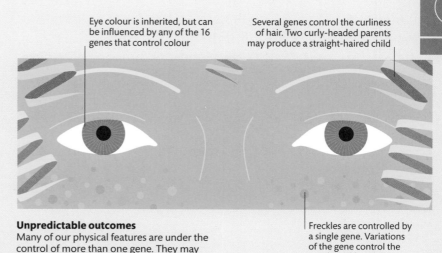

Eye colour is inherited, but can be influenced by any of the 16 genes that control colour

Several genes control the curliness of hair. Two curly-headed parents may produce a straight-haired child

Unpredictable outcomes
Many of our physical features are under the control of more than one gene. They may result in unexpected combinations.

Freckles are controlled by a single gene. Variations of the gene control the number of freckles

Unravelling DNA

Chromosomes help package DNA to fit into the nucleus. The DNA is wrapped around spool-like proteins that run through the centre of each chromosome. The helix is made of two strands of sugar phosphate linked together by a pair of bases. The bases always form the same pairs, but the sequences of bases along the strand are specific to the proteins they will eventually produce.

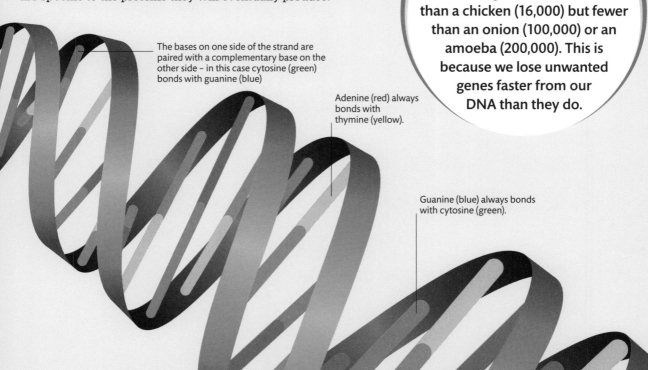

The bases on one side of the strand are paired with a complementary base on the other side – in this case cytosine (green) bonds with guanine (blue)

Adenine (red) always bonds with thymine (yellow).

Guanine (blue) always bonds with cytosine (green).

DO HUMANS HAVE THE MOST GENES?

Humans have a relatively low number of genes. We have more than a chicken (16,000) but fewer than an onion (100,000) or an amoeba (200,000). This is because we lose unwanted genes faster from our DNA than they do.

How cells multiply

We all start life as a single cell, so to develop specific tissues and organs and enable our body to grow, our cells need to multiply. Even as adults, cells need to be replaced because they get damaged or complete their life cycle. There are two processes by which this happens – mitosis and meiosis.

OUT OF CONTROL

Many cancers occur when a mutant cell begins to multiply rapidly. This is because the cell can override the usual checks during mitosis, enabling it to replicate itself more quickly than surrounding cells and take up more of the available oxygen and nutrients.

Cancerous cell

Wear and tear
Mitosis happens whenever new cells are needed. Some cells, such as neurons, are rarely replaced, but others, such as those lining the gut or tastebuds, undergo mitosis every few days.

1 Resting
The parent cell gets ready for mitosis by checking its DNA for damage and making any repairs needed.

Cell

Nucleus

Four of cell's 46 chromosomes

6 Offspring
Two daughter cells are formed, each containing a nucleus with an exact copy of the DNA from the parent cell.

Mitosis

Every cell enters a phase in its life cycle called mitosis. During mitosis, the cell's DNA is duplicated and then divides equally to form two identical nuclei, each containing the exact same DNA as the original parent cell. The cell then divides up its cytoplasm and organelles to form two daughter cells, each containing a single nucleus. There are a number of checkpoints throughout the DNA replication and division processes to repair any damaged DNA, which could lead to permanent mutations and disease.

2 Preparation
Each chromosome in the parent cell makes an exact copy of itself prior to entering mitosis. The copies join at a region called the centromere.

Centromere

5 Splitting
A nuclear membrane forms around each group of chromosomes and the cell membrane starts to pull apart to form two cells.

4 Separation
The chromosomes split at their attachment point (centromere) and each half is pulled to an opposite end of the cell.

3 Lining up
Each of the doubled chromosomes attaches to special fibres, which help to line them up in the middle of the cell.

Fibre

Centromere

1 Preparation
Each of the cell's chromosomes duplicates and the copies join together at the centromere.

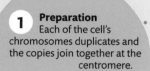

Cell
Nucleus
Chromosome
Centromere

2 Pairing and crossover
Chromosomes with similar lengths and centromere locations line up with one another and undergo gene swapping.

3 First separation
The chromosomes line up and, just like in mitosis, are pulled to opposite ends of the cell along special fibres.

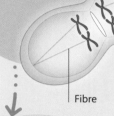

Fibre

Gene swapping
Meiosis features a unique process that shuffles the DNA passing into the daughter cells. DNA is exchanged between the chromosomes, which creates new combinations of DNA. Some new combinations may be beneficial.

6 Four offspring
Four cells are produced, each with half the number of chromosomes of the original parent cell, and each genetically unique.

5 Second separation
The chromosomes line up along the midline of each cell and are pulled apart so that each new cell receives half of the chomosome pair.

4 Two offspring
The cell divides, and two cells containing half the chromosomes are formed. Each is genetically distinct from each other and from the parent cell.

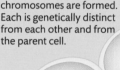

Meiosis

Egg and sperm cells are produced through a specialized type of cell division known as meiosis. The aim is to reduce the number of chromosomes from the parent cell by half so that when an egg and sperm fuse during fertilization, the new cell has a full complement of 46 chromosomes. Meiosis produces four daughter cells that are each genetically different to the parent cell. It is the process of gene swapping during meiosis that introduces the genetic diversity that helps make each of us unique individuals.

DOWN'S SYNDROME

Sometimes mistakes can happen during meiosis. Down's syndrome is caused by an extra copy of chromosome 21 in some or all of the body's cells. This usually happens when the chromosome doesn't separate properly during the meiosis of an egg or sperm cell – a condition known as trisomy 21. Having an extra chromosome means that some genes are overexpressed by the cell, which can cause problems in how it functions.

The extra 310 genes can result in overproduction of some proteins.

THREE COPIES OF CHROMOSOME 21

How genes work

If our DNA is the body's recipe book, then a gene within that DNA is equivalent to a single recipe in the book; it is the instructions for building a single chemical or protein. It's estimated that humans have around 20,000 genes that code for different proteins.

Genetic blueprint

To translate a gene into a protein, the DNA is first copied (transcribed) in the nucleus of a cell by enzymes, forming a strand of messenger RNA (mRNA). The cell will only copy those genes that it needs, not the entire DNA sequence. The mRNA then travels outside the nucleus where it can be translated into a chain of amino acids, which will build the protein.

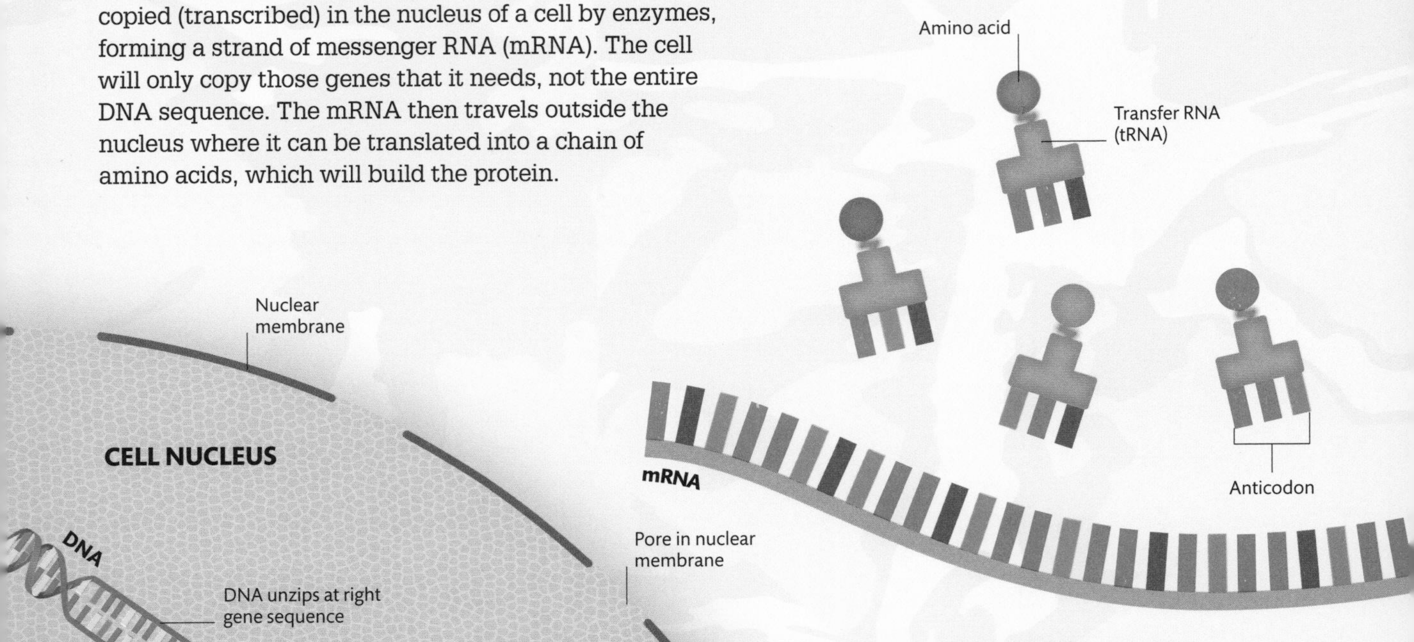

Amino acid

Transfer RNA (tRNA)

Anticodon

Nuclear membrane

CELL NUCLEUS

mRNA

DNA

Pore in nuclear membrane

DNA unzips at right gene sequence

RNA polymerase enzyme builds new strand of mRNA

mRNA contains matching base pairs to DNA strand

1 **Starting translation**
The newly made mRNA travels to a protein-building unit called a ribosome, to which it attaches. There, it attracts molecules of transfer RNA (tRNA), each of which has an amino acid attached to it.

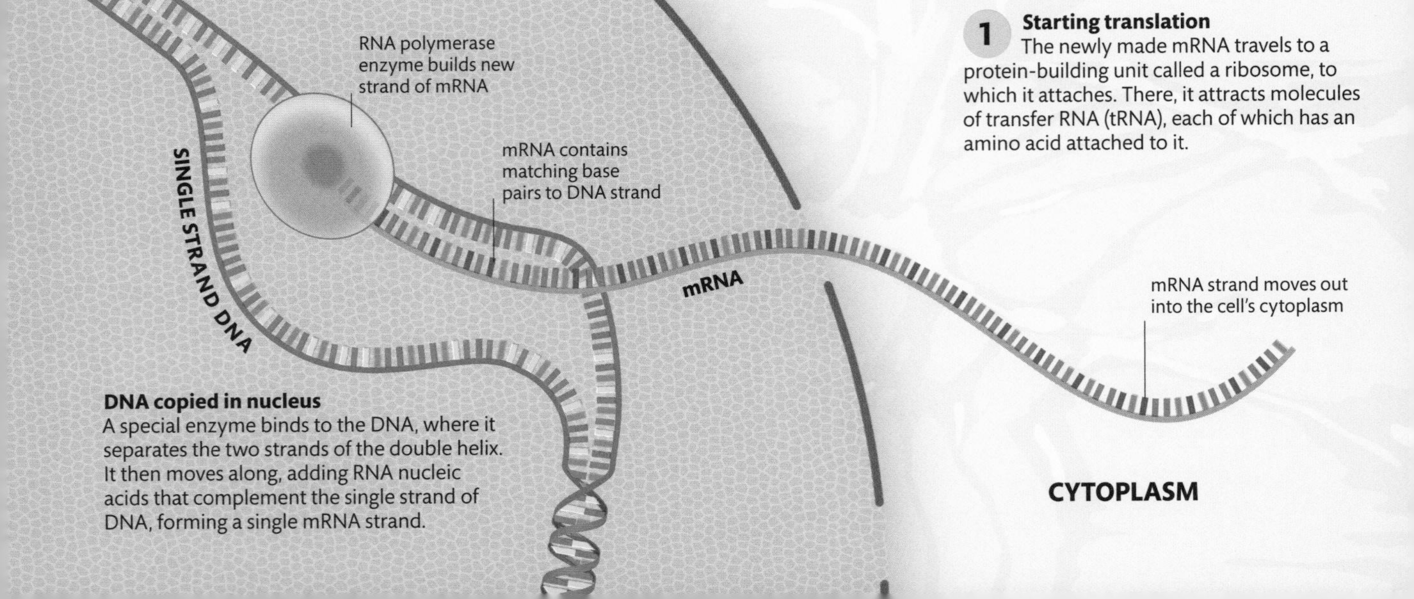

SINGLE STRAND DNA

mRNA

mRNA strand moves out into the cell's cytoplasm

DNA copied in nucleus

A special enzyme binds to the DNA, where it separates the two strands of the double helix. It then moves along, adding RNA nucleic acids that complement the single strand of DNA, forming a single mRNA strand.

CYTOPLASM

4 Amino acids folded into proteins

When the ribosome reaches a stop codon at the end of the mRNA strand, the long chain of amino acids is complete. The order of the amino acids determines how the chain folds up into a protein.

CHAIN FOLDED INTO PROTEIN

Chain of amino acids builds as ribosome moves along mRNA strand

Making proteins

Every three bases in the mRNA is known as a codon and each codon specifies a particular amino acid. There are 21 different amino acids and a single protein may be made up of a chain of hundreds of these amino acids.

2 Ribosome attaches amino acids

As the ribosome moves along the mRNA strand, the tRNA molecules attach to the mRNA in a specific order. This order is determined by the matching up of codons – a sequence of three nucleic acid bases on the mRNA strand – and their complementary three bases – called anticodons – on the tRNA molecule.

3 Building a chain

The amino acid detaches from the tRNA molecule and is joined to the previous amino acid with a peptide bond, forming a chain.

tRNA, once it has dropped off its animo acid, floats off into cytoplasm

RIBOSOME

Codon

LOST IN TRANSLATION

Gene mutations can cause changes in the amino acid sequence. A single mutation in the 402nd base of the gene that codes for the hair protein keratin causes the amino acid lysine to be put in place of glutamate. This changes the shape of the keratin, making the hair look beaded.

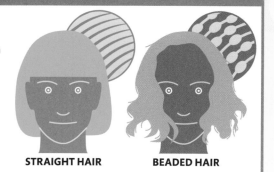

STRAIGHT HAIR   **BEADED HAIR**

WHAT HAPPENS TO mRNA AFTER TRANSLATION?

A strand of mRNA may be translated into a protein many times before it eventually degrades within the cell.

How genes make different cells

DNA contains all of the blueprints for life, but cells pick and choose only the plans (genes) they need. These genes are used by the cell to build the proteins and molecules that not only define what the cell looks like, but what it does within the body.

HOW DO CELLS KNOW WHAT TO DO?

The chemical environment around the cell, or signals from other cells, tell it that it is part of a particular tissue or organ, or in a certain stage of development.

Gene expression

Each cell uses, or "expresses", only a fraction of its genes. As it becomes more specialized, more genes are switched off. This process is highly regulated and happens in specific order, usually when the DNA is being transcribed to RNA (see pp.20–21).

1 Regulation
Transcription of a required gene is controlled by a series of genes that sit in front of it. These include regulator, promoter, and operator genes. The gene won't be transcribed until conditions are right.

REGULATOR PROTEIN   PROMOTER   OPERATOR

Gene to be transcribed (copied to RNA)

REGULATOR   GENE SEQUENCE

2 Repressor protein
If a repressor protein is blocking the gene, transcription can't take place. The gene can only be turned on when a change in the environment removes the repressor protein.

RNA POLYMERASE

REPRESSOR

Repressor protein prevents polymerase binding to DNA

Activator protein

Polymerase can now bind to the DNA and start transcription

3 Activation
When an activator protein binds to the regulator protein and there are no repressor proteins blocking the gene, transcription can start.

RNA POLYMERASE

On or off?

Embryonic cells start out as stem cells – cells with the ability to turn into different cell types. Stem cells initially have the same set of genes switched on and they simply keep growing and dividing to produce more cells. As an embryo develops, it needs its cells to specialize and organize into tissues and eventually organs. So when signalled, the cells start shutting off some genes and switching on others to turn into a specific type of cell.

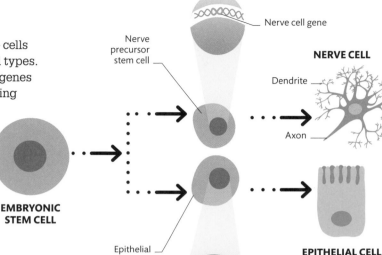

Nerve cell gene

Nerve precursor stem cell

NERVE CELL

Dendrite

Axon

EMBRYONIC STEM CELL

Epithelial precursor stem cell

EPITHELIAL CELL

Epithelial cell gene

Making a difference

As an embryo is developing, a stem cell destined to become a nerve cell will turn on the genes needed to grow dendrites and an axon, whereas another stem cell might activate different genes to become an epithelial (skin) cell.

Housekeeping proteins

Some proteins, such as DNA repair proteins or enzymes needed for metabolism, are called housekeeping proteins, because they are essential to the basic functioning of all cells. Many are enzymes, while others add structure to cells or help transport substances in and out of cells. The genes for these proteins are always turned on.

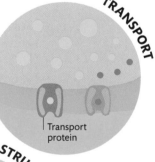

TRANSPORT

Transport protein

On the move
Special proteins are needed to move materials around the body or help them cross cell membranes.

STRUCTURE

Structural protein

Enzyme

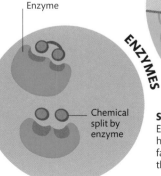

Chemical split by enzyme

ENZYMES

Speeding things up
Enzymes are proteins that help chemical reactions go faster, such as those used in the breakdown of food.

Providing support
Structural proteins are found in all cells. They give the cell its shape and hold the organelles in place.

BOY OR GIRL?

At 6 weeks, an embryo has all the internal organs needed to be either male or female. If it is genetically a male embryo, a gene on the Y chromosome will turn on at this stage and produce the hormones that develop the male reproductive organs and cause the female organs to degenerate. The reason why men have seemingly pointless nipples is that these are also formed in the first 6 weeks, but their further development depends on whether they are in a male or female hormonal environment.

Adult stem cells

Adult stem cells have been found in the brain, bone marrow, blood vessels, skeletal muscles, skin, teeth, heart, gut, liver, ovaries, and testes. These cells can sit inactive for a long time until they are called into action to replace cells or repair damage, when they begin to divide and specialize. Researchers can manipulate these cells to become specific cell types that can then be used to grow new tissues and organs.

WHERE DO ADULT STEM CELLS COME FROM?

This is currently being investigated, but one theory is that some embryonic stem cells remain in various tissues after development.

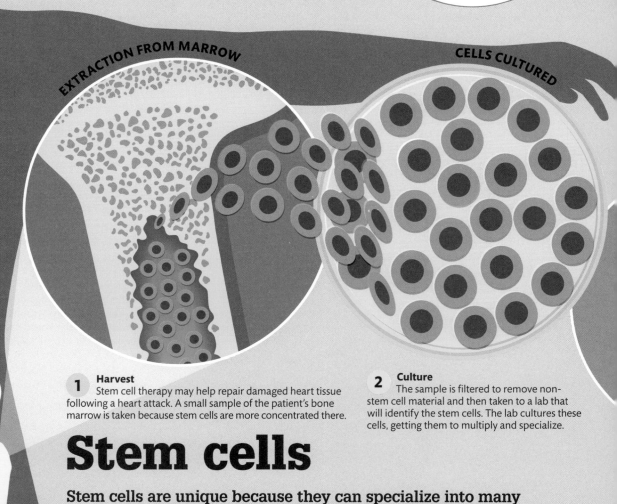

EXTRACTION FROM MARROW

CELLS CULTURED

1 Harvest
Stem cell therapy may help repair damaged heart tissue following a heart attack. A small sample of the patient's bone marrow is taken because stem cells are more concentrated there.

2 Culture
The sample is filtered to remove non-stem cell material and then taken to a lab that will identify the stem cells. The lab cultures these cells, getting them to multiply and specialize.

Stem cells

Stem cells are unique because they can specialize into many different types of cells. Stem cells are the foundation for the body's repair mechanisms, which makes them potentially useful in helping repair damage in the body.

ADULT OR EMBRYONIC CELLS?

Most research has focused on embryonic stem cells as they are capable of developing into any cell type. This research is controversial, as embryos – created using donor eggs and sperm – are grown specifically for the purpose of harvesting the cells. The focus has since turned to adult stem cells, as they can now be turned into a wider range of cells than originally thought.

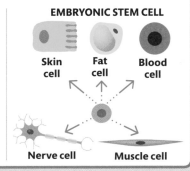

UNTREATED ADULT STEM CELL

Red blood cell

White blood cell

Platelet

EMBRYONIC STEM CELL

Skin cell

Fat cell

Blood cell

Nerve cell

Muscle cell

Engineering tissues

Researchers have found that the physical structure of the supporting matrix (scaffold) used to grow stem cells is critical to the way they grow and specialize.

1 Taking shape
To repair the eye's cornea, stem cells are extracted from a healthy tissue (the cornea of the unaffected eye) and grown on a dome-shaped mesh.

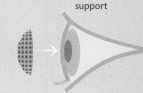

Stem cells

Mesh support

2 Transplant
The damaged cells on the cornea of the eye are removed and replaced with the mesh structure. After several weeks, the mesh dissolves leaving the grafted cells, which have restored the patient's sight.

Potential uses of stem cells
Stem cell research has improved our understanding of embryonic development and the natural repair mechanisms in the body. The most active area of research is their use growing replacement organs and reconnecting the spinal cord so that paralysed people can walk again.

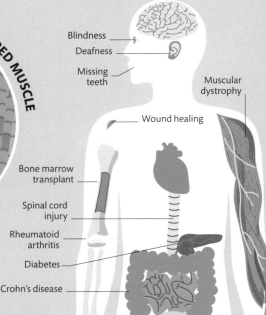

Blindness

Deafness

Missing teeth

Muscular dystrophy

Wound healing

Bone marrow transplant

Spinal cord injury

Rheumatoid arthritis

Diabetes

Crohn's disease

Osteoarthritis

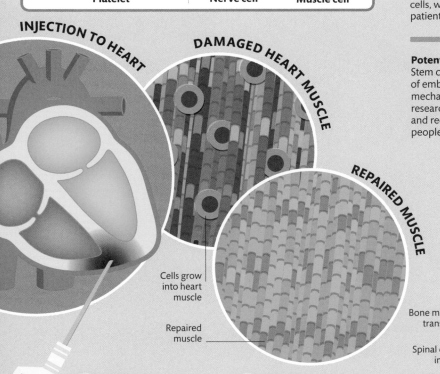

INJECTION TO HEART

DAMAGED HEART MUSCLE

REPAIRED MUSCLE

Cells grow into heart muscle

Repaired muscle

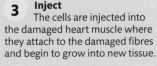

3 Inject
The cells are injected into the damaged heart muscle where they attach to the damaged fibres and begin to grow into new tissue.

4 Repair
After several weeks, the damaged heart muscle is rejuvenated. This process also reduces scarring that would restrict the heart's movement.

Environmental assault

Each of our cells is inundated daily by chemicals and energy that can cause damage to our DNA. Solar radiation (UV), environmental toxins, and even the chemicals produced through our own cellular processes can cause changes to our DNA that affect how it works, including how it can be copied or how it produces proteins. If this damage becomes a permanent change in the DNA, it is called a mutation.

20,000
THE NUMBER OF **DAMAGED BASES REMOVED** AND **REPLACED** IN EVERY CELL **EVERY DAY**

CAN THE DAMAGE ALWAYS BE REPAIRED?

Our ability to repair DNA diminishes as we get older. Damage starts to accumulate and this is thought to be one of the main reasons behind ageing.

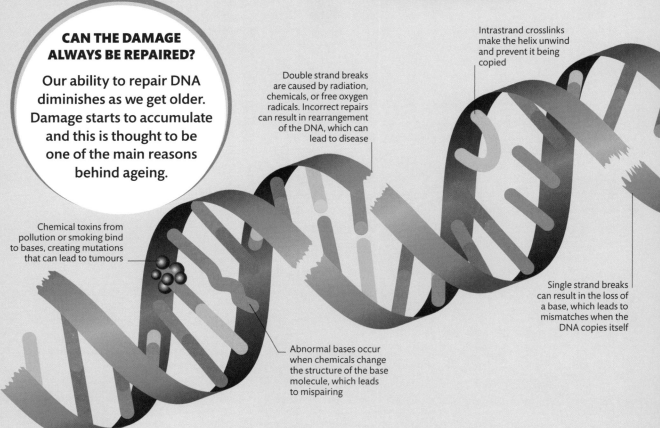

Intrastrand crosslinks make the helix unwind and prevent it being copied

Double strand breaks are caused by radiation, chemicals, or free oxygen radicals. Incorrect repairs can result in rearrangement of the DNA, which can lead to disease

Chemical toxins from pollution or smoking bind to bases, creating mutations that can lead to tumours

Single strand breaks can result in the loss of a base, which leads to mismatches when the DNA copies itself

Abnormal bases occur when chemicals change the structure of the base molecule, which leads to mispairing

When DNA goes wrong

Every day, the DNA in cells is damaged – whether by natural processes or environmental factors. This damage can affect DNA copying or how specific genes function and if it can't be repaired, or is repaired incorrectly, it can lead to disease.

UNDER ATTACK

This DNA strand is shown under many kinds of stress. However, some types of DNA damage can be used to advantage. Many chemotherapy drugs are designed to cause damage to the DNA in cancerous cells. Cisplatin, for example, forms crosslinks in the DNA, which triggers cell death. Unfortunately it also causes damage in normal healthy cells.

Interstrand crosslinks between the same bases halt DNA copying because they prevent the strands from unzipping

Base mismatches occur when an extra base has been added or one has been skipped in the replication process

The insertion or deletion of bases means that when the code is being read during copying, the wrong proteins will be produced

Gene therapy

When DNA damage causes a mutation, it can stop a gene from working properly and result in disease. While drugs might help treat the symptoms of the disease, they can't solve the underlying genetic problem. Gene therapy is an experimental method that's exploring ways to fix the defective gene.

REPAIRING DNA

Cells have built-in safety systems that help to identify and repair damage to their DNA. These systems are constantly active and if they are unable to fix the damage quickly, they will stop the cell cycle temporarily so they can take some extra time to work on it. If it's not repairable, they will trigger the death of the cell by apoptosis (see p.15).

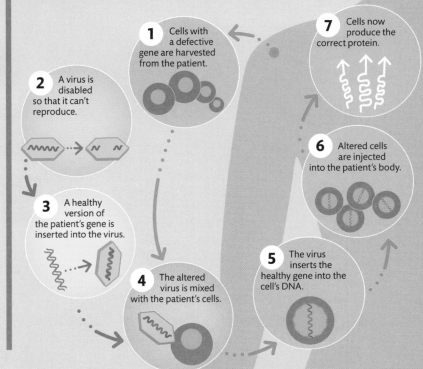

1 Cells with a defective gene are harvested from the patient.

2 A virus is disabled so that it can't reproduce.

3 A healthy version of the patient's gene is inserted into the virus.

4 The altered virus is mixed with the patient's cells.

5 The virus inserts the healthy gene into the cell's DNA.

6 Altered cells are injected into the patient's body.

7 Cells now produce the correct protein.

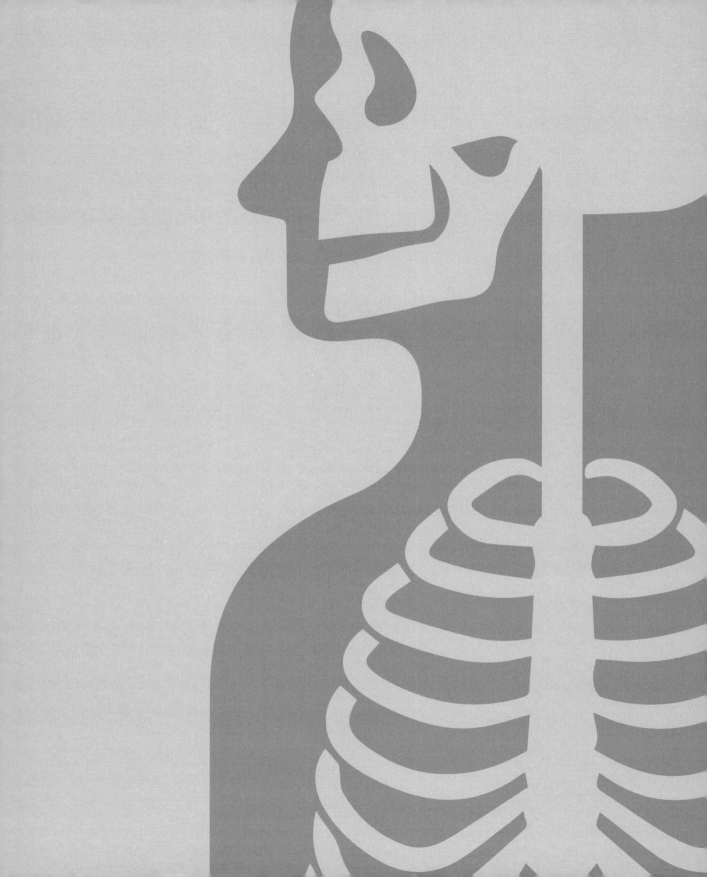

HOLDING IT
TOGETHER

Skin deep

The skin is the largest organ of the human body. It protects us from physical damage, dehydration, overhydration, and infection, but also regulates body temperature, makes vitamin D, and has an extraordinary array of special nerve endings (see pp.74–75).

Keeping cool and staying warm

Humans have adapted to survive in the heat of the tropics, the cold of the arctic, and the temperate climates in between. Although we have lost most of our body hair and rely on clothes to keep us warm, even fine body hair plays a role in controlling body temperature. In hot weather, it is vital to drink plenty of water to replace the sweat that helps to keep us cool.

Hot-weather skin

Each day, the skin's 3 million sweat glands secrete 1 litre (1¾ pints) of sweat, or up to 10 litres (16 pints) daily in extreme conditions. Evaporation of sweat takes the heat energy away from the body. Ring-shaped muscles around the blood vessels also help by diverting blood to the skin, which lets heat escape from deep in the body.

Cold-weather skin

In cold weather, the skin goes into heat-retention mode. Tiny muscles stand our body hairs upright, trapping warmth close to the skin. Meanwhile, the capillary network muscles stop warm blood flowing into the skin's surface layers.

Hair reclines to release the heat around it

Sweat droplets evaporate, taking heat away with them

Heat rises to the surface of the skin from the capillary network

Hair stands up to trap the heat around it

Hair erector muscle contracts

The skin rises into a "goose bump" around the hair

Sweat production stops

CAPILLARY NETWORK

SWEAT GLAND

Fat in the lowest layer of the skin retains heat

BLOODSTREAM

Muscle in the capillary network relaxes, shunting blood to the outer layers of the skin

Hair erector muscle relaxes, allowing the hair to flatten

Capillary muscle contracts, reducing the flow of blood to the outer layers of the skin

Defensive barriers

The skin is made up of three layers, each of which plays a vital role in our survival. The upper layer, called the epidermis, is an ever-regenerating defence system (see pp.32–33) that has its roots in the middle layer, called the dermis. The final layer is the hypodermis – a cushion of fat that keeps us warm, protects our bones, and keeps us supplied with energy (see pp.158–59).

THE **SKIN** OF AN **AVERAGE ADULT** MEASURES **2 SQ M (21 SQ FT)** IN **AREA**

DO GOOSE BUMPS REALLY HELP?

Goose bumps do help us retain heat in cold weather. However, they were much more effective millions of years ago, when we were covered in thick hair. The thicker the hair, the more heat is trapped when the hair stands on end.

Microbe   Sebum

Anti-bacterial oil
Glands secrete an oil called sebum into hair follicles to condition the hair and waterproof the skin. Sebum also suppresses the growth of bacteria and fungi.

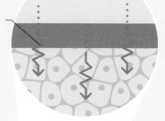

Ultraviolet light

Ultraviolet light protection
The skin uses ultraviolet light to synthesize vitamin D – but too much ultraviolet light can cause skin cancer. A skin pigment called melanin helps maintain a balance between the two (see pp.32–33).

Sebaceous gland secretes sebum

Ever-regenerating epidermal cells

NICOTINE PATCH

EPIDERMIS

HAIR SHAFT

DERMIS

Nicotine reaches the bloodstream

One of the skin's many types of nerve endings (see pp.74–75)

Letting things pass
Although skin is a barrier, it is selectively permeable, letting through drugs, such as nicotine and morphine, from patches applied to the skin's surface. Various creams, such as sunblock, moisturizer, and antiseptic cream can also cross the barrier.

HAIR BULB

The epidermis stretches all the way under the hair bulb

HYPODERMIS

Outer defences

The skin is the frontier between us and the outside world – a boundary at which enemies are fought and friends let in. Key features of its defences are a self-renewing outer layer and a pigment that shields us from ultraviolet light.

The self-renewing layer

The epidermis is a conveyor belt of cells, which are constantly forming at is base – the basal layer – and travelling upwards to the surface. As they move, they lose their nucleus, flatten, and fill with a tough protein called keratin, and so form a protective, outer layer. This layer is constantly being worn away and replaced by new, upthrusting cells. Each cell dies by the time it reaches the surface. These dead cells then fall off and contribute to the dust in our houses.

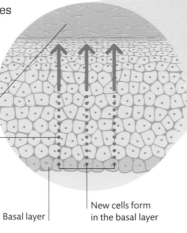

Dead cell flakes off

Cells travel up through the epidermis

Basal layer

New cells form in the basal layer

Transparent defence
Because the epidermis sheds its cells, tattoos have to be inscribed beneath it, on the dermis. Luckily, the epidermis is transparent, so tattoos can be seen through it.

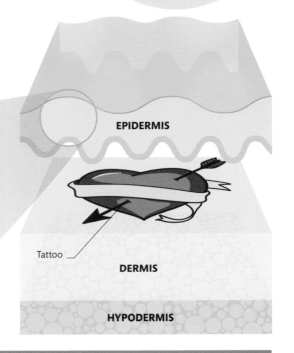

EPIDERMIS

Tattoo

DERMIS

HYPODERMIS

Scaffolding

Beneath the epidermis lies the dermis, a thick layer that gives the skin its strength and flexibility. It contains the skin's nerve endings, sweat glands, oil glands, hair roots, and blood vessels. It is made primarily of collagen and elastin fibres, which form a kind of scaffolding that enables the skin to stretch and contract in response to pressure.

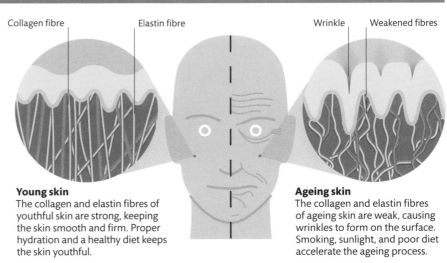

Collagen fibre

Elastin fibre

Wrinkle

Weakened fibres

Young skin
The collagen and elastin fibres of youthful skin are strong, keeping the skin smooth and firm. Proper hydration and a healthy diet keeps the skin youthful.

Ageing skin
The collagen and elastin fibres of ageing skin are weak, causing wrinkles to form on the surface. Smoking, sunlight, and poor diet accelerate the ageing process.

Skin colour

One of the skin's many functions is to make vitamin D, which it does by harnessing ultraviolet (UV) light from the Sun. However, UV light is also very dangerous (it can cause skin cancer), so we also need protecting from it. As protection, the skin produces melanin – a pigment that serves as a Sun shield, and so determines skin colour.

FRECKLES ARE CAUSED BY **MELANOCYTES CLUMPING TOGETHER**

Dark skin

At the equator, the Sun's rays strike the Earth almost vertically, and so with great intensity. This means that people living near the equator have a great need of UV protection. To provide this, the skin produces large amounts melanin – which results in dark skin.

Intense rays of UV light

5 UV shield
Melanosomes break apart, spreading melanin across the skin. This forms a shield against UV rays.

2 Dendrites
Melanocytes have finger-like extensions called dendrites. Each of these touches around 35 neighbouring cells.

4 Absorption
Melanosomes are absorbed by neighbouring skin cells.

1 Melanocytes
Melanin is produced by special cells called melanocytes. These are embedded in the base of the epidermis.

3 Melanosomes
Melanin moves along the dendrites in packets called melanosomes.

Melanosome

Dendrite

Melanocyte

Basal layer

Pale skin

North and south of the equator, the Sun's rays hit the Earth at increasingly shallow angles. The shallower the angle, the less intense the light, and the less the need for UV protection. In response, the skin produces smaller amounts of melanin – which results in pale skin.

Mild rays of UV light

Dendrite

3 Weaker shield
The weaker melanin shield is sufficient against weaker UV rays.

1 Melanocytes
In pale skin, the melanocytes are less active, and have fewer dendrites.

2 Paler melanosomes
Melanosomes are paler and taken up by fewer surrounding cells.

Melanocyte

Melanosome

The extremities

Hair and nails are both made of a tough, fibrous protein called keratin. Nails strengthen and protect the tips of your fingers and toes, while hair reduces heat loss from the body to help keep you warm.

Hair colour, thickness, and curliness

Each hair has a spongy core (medulla) and a middle layer (cortex) of flexible protein chains that give it wave and bounce. An outer layer (cuticle) of scales reflects light so hair looks shiny, but if these are damaged, hair looks dull. The colour, curliness, thickness, and length of your hair are determined by the size and shape of your follicles (in which they grow), and the types of pigment they produce.

WHY DOES HAIR LENGTH VARY?

Scalp hair can grow for years, but hair found elsewhere on the body only grows for weeks or months. That's why body hair is usually short – it falls out before it can grow very long.

Thick, straight, and red
A mixture of pale and dark melanin produces hair that is gold, auburn or red. Large, round follicles produce thick hair. Thickness also depends on the number of active follicles present. Redheads tend to have relatively few follicles.

A large proportion of pheomelanin

A little eumelanin

Fine, straight, and blonde
Cells at the base of each follicle feed melanin pigments through to the root. Blonde hair contains a pale melanin pigment that is only present in the middle of the shaft (medulla). Small, round follicles produce straight, fine hair.

Medulla

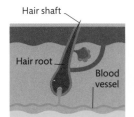

Cuticle

Pale melanin pigment, pheomelanin

Cortex

Scales

A little dark melanin, or eumelanin

Hair growth

Each hair follicle goes through around 25 cycles of hair growth during its lifespan. Each cycle has a growth stage when it lengthens, followed by a resting phase in which the hair remains the same length, starts to loosen, and falls out. After the resting phase, the follicle reactivates and starts to produce a new hair.

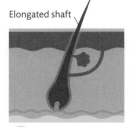

Hair shaft

Hair root

Blood vessel

Elongated shaft

Hair bulb

1 Early growth
The follicle activates, producing new cells within the hair root. These die and are pushed upwards to form the shaft.

2 Late growth
The shaft elongates over a period of 2–6 years. A longer growth period (more common in women) produces longer hair.

3 Resting
The follicle shrinks and the hair stops growing as the bulb pulls away from the root. This takes 3–6 weeks.

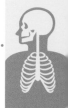

Thick, black, and curly
Dark hair contains black melanin pigment in both the cortex and the medulla, producing more depth of colour. Wavy hair grows from oval-shaped follicles. As follicles become flatter, hair curliness increases.

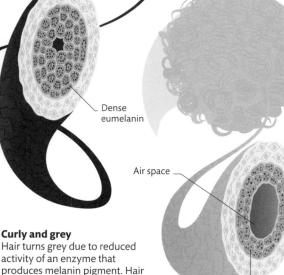

Dense eumelanin

Air space

Impoverished eumelanin

Curly and grey
Hair turns grey due to reduced activity of an enzyme that produces melanin pigment. Hair without melanin is snow white; hair with a little pigment appears grey.

Nails

Nails are transparent plates of keratin. They act as splints to stabilize the soft flesh of your fingertips, and improve your grip on small objects. Nails also contribute to the overall sensitivity of your fingertips. However, because they project from the body, nails are easily damaged.

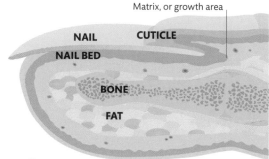

Matrix, or growth area

NAIL   **CUTICLE**

NAIL BED

BONE

FAT

How nails grow
Growing areas at the base and sides of each nail are protected by folds of skin called cuticles. Cells in the nail beds are among the most active in the body. They are constantly dividing, and nails grow up to 5 mm (⅕ in) per month.

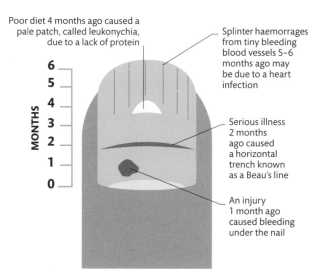

Poor diet 4 months ago caused a pale patch, called leukonychia, due to a lack of protein

Splinter haemorrages from tiny bleeding blood vessels 5–6 months ago may be due to a heart infection

Serious illness 2 months ago caused a horizontal trench known as a Beau's line

An injury 1 month ago caused bleeding under the nail

MONTHS

6
5
4
3
2
1
0

Diary of a nail
Because nails are non-essential structures, blood and nutrients are diverted away from the nail beds in times of deficiency. Nails are therefore a good indicator of your general health and diet. A doctor glances quickly at a patient's hands as the nails can indicate a number of illnesses.

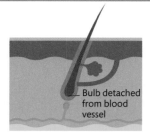

4 **Detachment**
The loose hair is shed naturally, or dislodged by brushing or combing. Sometimes it's pushed out by a new hair growing.

Bulb detached from blood vessel

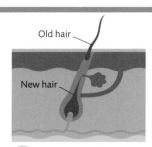

Old hair

New hair

5 **New growth**
The follicle starts its next cycle. With age, fewer follicles reactivate, so hair becomes thinner, recedes, and bald areas may appear.

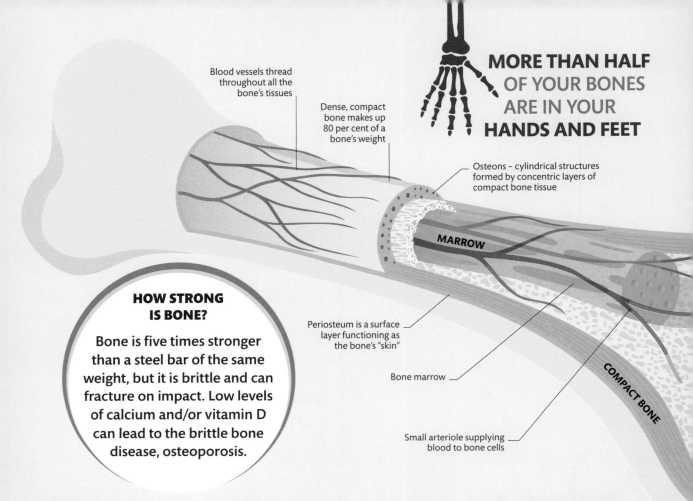

Blood vessels thread throughout all the bone's tissues

Dense, compact bone makes up 80 per cent of a bone's weight

MORE THAN HALF OF YOUR BONES ARE IN YOUR **HANDS AND FEET**

Osteons – cylindrical structures formed by concentric layers of compact bone tissue

MARROW

COMPACT BONE

Periosteum is a surface layer functioning as the bone's "skin"

Bone marrow

Small arteriole supplying blood to bone cells

HOW STRONG IS BONE?

Bone is five times stronger than a steel bar of the same weight, but it is brittle and can fracture on impact. Low levels of calcium and/or vitamin D can lead to the brittle bone disease, osteoporosis.

Pillars of support

Your skeleton is rather like a coat hanger on which your flesh is draped. As well as giving your body support and shape, your bones provide protection and, through their interaction with muscles, allow your body to move and adopt different poses.

Living tissue

Bone is a living tissue made of collagen protein fibres filled with minerals – calcium and phosphate – that give them rigidity. Bones contain 99 per cent of all the calcium in your body. Bone cells constantly replace old, worn out bone with new bone tissue. Blood vessels supply these cells with oxygen and nutrients. A surface layer of skinlike periosteum covers a shell of compact bone, which provides strength. Beneath this is a sponge-like network of struts that reduces the overall weight. Bone marrow in certain bones, including the ribs, breast bone, shoulder blades, and pelvis, has a special job – it produces new blood cells.

THE SMALLEST BONES

The stapes in the middle ear is the smallest named bone. You also have small, sesamoid bones (named after the sesame seeds they resemble) in long tendons at sites of pressure to prevent the tendons wearing away.

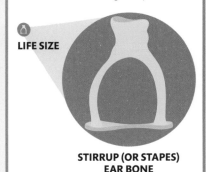

LIFE SIZE

STIRRUP (OR STAPES) EAR BONE

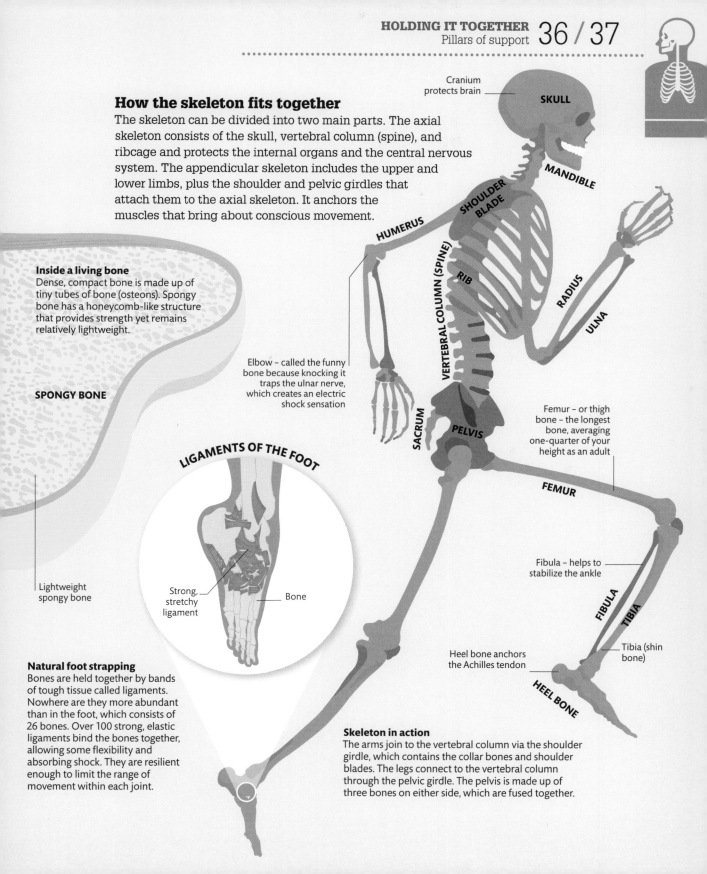

How the skeleton fits together

The skeleton can be divided into two main parts. The axial skeleton consists of the skull, vertebral column (spine), and ribcage and protects the internal organs and the central nervous system. The appendicular skeleton includes the upper and lower limbs, plus the shoulder and pelvic girdles that attach them to the axial skeleton. It anchors the muscles that bring about conscious movement.

Cranium protects brain

SKULL

MANDIBLE

SHOULDER BLADE

HUMERUS

VERTEBRAL COLUMN (SPINE)

RIB

RADIUS

ULNA

SACRUM

PELVIS

FEMUR

FIBULA

TIBIA

HEEL BONE

Inside a living bone

Dense, compact bone is made up of tiny tubes of bone (osteons). Spongy bone has a honeycomb-like structure that provides strength yet remains relatively lightweight.

SPONGY BONE

Elbow – called the funny bone because knocking it traps the ulnar nerve, which creates an electric shock sensation

Femur – or thigh bone – the longest bone, averaging one-quarter of your height as an adult

Fibula – helps to stabilize the ankle

Tibia (shin bone)

Heel bone anchors the Achilles tendon

Lightweight spongy bone

LIGAMENTS OF THE FOOT

Strong, stretchy ligament

Bone

Natural foot strapping

Bones are held together by bands of tough tissue called ligaments. Nowhere are they more abundant than in the foot, which consists of 26 bones. Over 100 strong, elastic ligaments bind the bones together, allowing some flexibility and absorbing shock. They are resilient enough to limit the range of movement within each joint.

Skeleton in action

The arms join to the vertebral column via the shoulder girdle, which contains the collar bones and shoulder blades. The legs connect to the vertebral column through the pelvic girdle. The pelvis is made up of three bones on either side, which are fused together.

Growing bones

If you are a healthy baby, you measure 46–56 cm (18–22 in) in length at birth. You grow rapidly during infancy, as your bones elongate. Bone growth slows during childhood, but speeds up again at puberty. Your bones stop growing at around 18 years of age, when you reach your final adult height.

NEWBORN BABY WEIGHT

An average newborn baby weighs 2.5–4.3 kg (5½–9½ lb). Babies normally lose weight in the first days after birth, but by 10 days, most babies have regained their birth weight and start to put on around 28 g (1 oz) per day.

How bones grow

Growth in height occurs at special growth plates at the ends of the long bones. Bone growth is controlled by growth hormone, with an additional growth spurt occurring in response to sex hormones at puberty (see pp.222–23). The cartilage growth plates fuse by adulthood, after which no further increases in height are possible.

Articular cartilage

Cartilage growth plate (epiphysis)

New bone formation (secondary ossification centre)

Cartilage growth plate (epiphysis)

Medullary cavity (marrow formation)

Compact bone

Cartilage

Articular cartilage

Developing periosteum

Developing spongy bone (primary ossification centre)

Compact bone

Spongy bone

Cartilage waiting to convert into bone

Medullary cavity containing bone marrow

Spongy bone

1 Embryo
Bones initially form from soft cartilage that acts as a scaffold on which minerals are laid down. Hardened bone starts forming when the fetus reaches 2–3 months of development in the womb.

2 Newborn baby
At birth, bones still consist mostly of cartilage, but there are active sites of bone formation (ossification). The first to develop is the primary ossification centre in the shaft, followed by those at the ends.

3 Child
Most of the bone shaft consists of hardened compact and spongy bone. Growth plates (epiphyses) at both ends allow growth in length. Bone is still soft and can bend on impact to form a greenstick fracture.

4 Teenager
At puberty, a surge in sex hormones causes a rapid growth spurt. Increases in height occur when new bone is laid down at the cartilage growth plates (epiphyses) to lengthen the bone shaft.

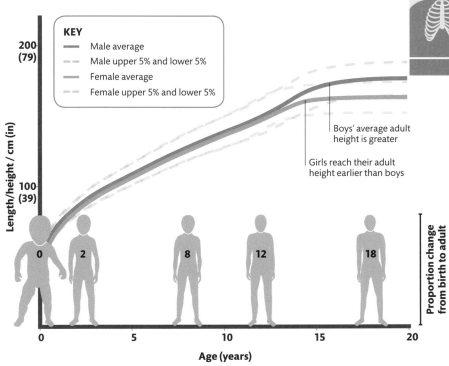

Boys' average adult height is greater

Girls reach their adult height earlier than boys

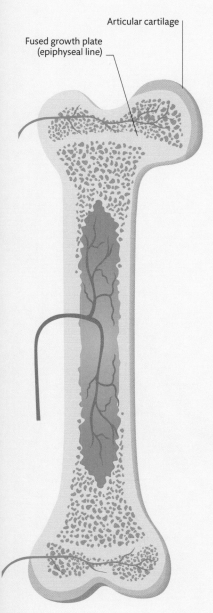

Articular cartilage

Fused growth plate (epiphyseal line)

5 **Adult**
After puberty, the cartilage growth plates are converted into bone (calcified) and fuse. This leave a hardened area called the epiphyseal line. Bones can still increase in diameter, but can no longer increase in length.

Growth patterns

A baby's head is one-quarter of his or her total body length. Changes in relative growth means that by the age of 2, that ratio is down to one-sixth. An adult's head is only one-eighth of body length. Girls enter puberty earlier than boys and reach their adult height around 16–17 years of age. Males only reach their final height between the ages of 19 and 21.

HOW TO CALCULATE YOUR FINAL HEIGHT

Assuming both parents are of normal stature, a child's potential adult height can be calculated as follows. Add father's height to mother's height. For a boy, add 13 cm (5 in) and for a girl deduct 13 cm (5 in). Then divide the total by two. Most children will have a final adult height within 10 cm (4 in) of this estimate.

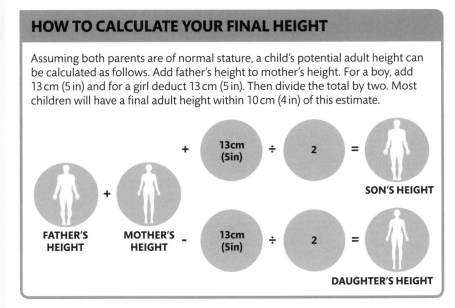

FATHER'S HEIGHT + MOTHER'S HEIGHT + 13cm (5in) ÷ 2 = SON'S HEIGHT

FATHER'S HEIGHT + MOTHER'S HEIGHT − 13cm (5in) ÷ 2 = DAUGHTER'S HEIGHT

Flexibility

Your joints allow you to move your body and manipulate objects. Movements can be small and controlled, such as when writing your name, or large and powerful, such as when throwing a ball.

Joint structure

A joint forms where two bones come into close contact. Some joints are fixed, with the bones locked together, such as the suture joints in an adult skull. Some joints have a limited range of movement, such as the elbow, while others can move more freely, such as the shoulder.

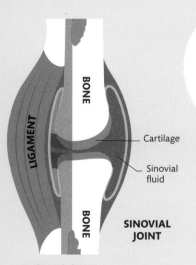

BONE

LIGAMENT

Cartilage

Sinovial fluid

BONE

SINOVIAL JOINT

Inside a joint

The bone ends within a mobile joint are coated with slippery cartilage and oiled with synovial fluid to reduce friction. These synovial joints are held together by bands of connective tissue, called ligaments. Some joints, such as the knee, also have internal stabilizing ligaments to stop the bones sliding apart while bending.

Ellipsoidal

These complex joints involve a bone with a rounded, convex end fitting into a bone with a hollow or concave shape. This allows a variety of movements, including sideways tilting, but not rotation.

Ball and socket

Found in the shoulders and hips, this type of joint allows the widest range of movement, including rotation. The shoulder joint is the most mobile joint in the body.

Gliding

These allow one bone to slide over another in any direction within one plane. Gliding joints allow the vertebrae to slide over each other when you flex your back. They are also found in your feet and hands.

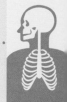

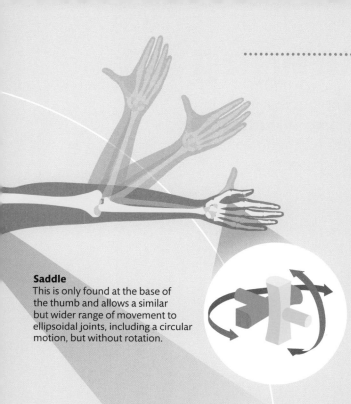

Types of joints

Although your body as a whole moves in complex ways, each individual joint has only a limited range of movement. A few joints have a very limited amount of movement so that they can absorb shock, such as where the two long bones in your lower leg (tibia and fibula) meet or some of the joints in the feet. The temporomandibular joints (see pp.44–45) between your jawbone and each side of the skull are unusual in that they each contain a disc of cartilage that allows the jaw to glide from side to side and protrude forwards and backwards during chewing and grinding your food.

Saddle
This is only found at the base of the thumb and allows a similar but wider range of movement to ellipsoidal joints, including a circular motion, but without rotation.

Pivot
This allows one bone to rotate around another, for example when you move your forearm to twist your palm to face up or down. A pivot joint in your neck allows your head to turn from side to side.

THE **SMALLEST JOINTS** ARE FOUND BETWEEN THE **THREE TINY BONES OF THE MIDDLE EAR** THAT HELP TO **TRANSMIT SOUND WAVES** TO THE **INNER EAR**

Hinge
This type of joint mainly allows movement in one plane, rather like a door opening and closing. Good examples are found in the elbow and knee.

DOUBLE-JOINTED PEOPLE

People who are said to be double-jointed have the same number of joints as everyone else, but their joints have a wider than normal range of movement. This trait is usually due to inheriting unusually elastic ligaments or a gene that codes for the production of a weaker type of collagen (a protein found in ligaments and other connective tissues).

Biting and chewing

Humans struggle to swallow large pieces of food so your teeth break down food as part of the first stage of digestion. Teeth also play a role in speech – it would be difficult to make the sound "tutt" without any teeth, for example.

From baby to adult

Your teeth are all present at birth as tiny buds deep within each jaw bone. The first "milk" teeth need to be small to fit within an infant's mouth. These teeth are shed during childhood as the mouth enlarges, leaving more room for adult-sized teeth.

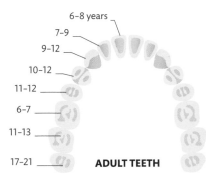

6–12 months
10–19
16–23
9–18
23–33

BABY TEETH

Eruption of milk teeth
The 20 milk teeth usually start to appear between the ages of 6 months and 3 years, though some infants have to wait a year.

6–8 years
7–9
9–12
10–12
11–12
6–7
11–13
17–21

ADULT TEETH

Eruption of adult teeth
The 32 adult teeth appear between the ages of 6 and 20 years and should last for the rest of your life – even if you live to be 100.

MUCH LIKE A **FINGERPRINT,** EACH PERSON HAS A **UNIQUE BITE IMPRESSION**

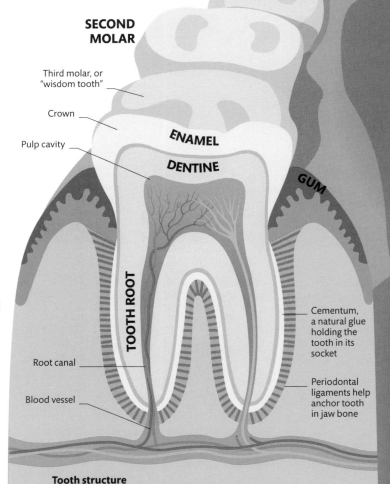

FIRST MOLAR
SECOND PREMOLAR
FIRST PREMOLAR
CANINE
SECOND INCISOR

SECOND MOLAR

Third molar, or "wisdom tooth"

Crown

Pulp cavity

ENAMEL

DENTINE

GUM

TOOTH ROOT

Root canal

Blood vessel

Cementum, a natural glue holding the tooth in its socket

Periodontal ligaments help anchor tooth in jaw bone

Tooth structure
Each tooth has a crown, above the gum, which is coated in hard enamel. This protects the softer dentine forming the tooth root. The central pulp cavity contains blood vessels and nerves.

FIRST INCISOR

SECOND INCISOR

CANINE

WHAT ARE WISDOM TEETH?

The last set of molars usually appear between the ages of 17 and 25. It is thought that they are called wisdom teeth because they appear after childhood.

Infection

Tooth enamel is the hardest substance in the body, but readily dissolves in acid, exposing the underlying parts of the tooth to bacteria and infection. Acid can come from some foods, juices, and fizzy drinks, or from bacterial plaque, which breaks down sugar to form lactic acid.

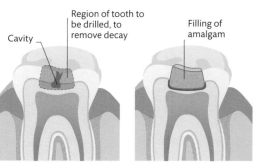

Cavity

Region of tooth to be drilled, to remove decay

Filling of amalgam

DECAYING TOOTH

TOOTH WITH FILLING

Decay and filling
When the hard enamel dissolves, it allows infection to rot the softer dentine beneath. A cavity forms as the weakened enamel overhead collapses.

Different types

Your teeth differ in shape and size depending on their use. Sharp-edged incisors cut and bite, canines tear, and molars and premolars have flattened, ridged surfaces that chew and grind food into tiny pieces.

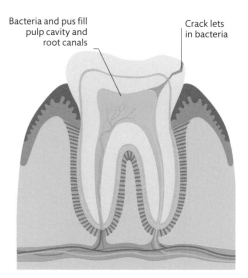

Bacteria and pus fill pulp cavity and root canals

Crack lets in bacteria

TOOTH WITH AN ABSCESS

Abscess
If bacteria reaches the pulp cavity, it may set up an infection in a place that is difficult for the immune system to tackle, and lead to an abscess that can spread to the jaw bone.

ARE YOU A GRINDER?

One in twelve people grind their teeth while asleep, and as many as one in five clench their jaws while awake. Known as bruxism, this weakens your teeth. You could be a grinder if your teeth look worn down, flattened or chipped, if your teeth are increasingly sensitive, or if you wake with jaw pain, a tightness in your jaw muscles, earache, or a dull headache – especially if you also chewed the inside of your cheeks. Worn down teeth may be reshaped with crowns.

FLATTENED TEETH

AFTER TREATMENT

The grinder

Your jaws are powered by strong muscles that produce considerable pressure as you cut and grind food with your teeth. The lower jaw can withstand these forces because it is the hardest bone in your body.

How we chew

Chewing is a complex motion in which the temporalis and masseter muscles control movement of the jaw back and forth, up and down, and side to side. This grinds food between the back molars like a pestle and mortar. The flexibility of the joints in our jaws allow us to slide effortlessly between chewing movements, depending on what we are eating.

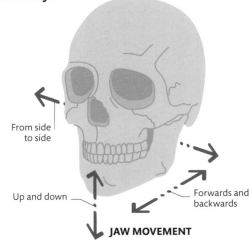

From side to side

Up and down

Forwards and backwards

JAW MOVEMENT

WHEN WE ATE LEAVES

Once, our primitive ancestors had smaller skulls and a chewier diet, rather like today's gorilla, pictured. Their powerful jaw muscles were anchored by a tall, sagittal crest along the top of the skull. This acted in a similar way to the breastbone of a bird, which anchors its giant flight muscles.

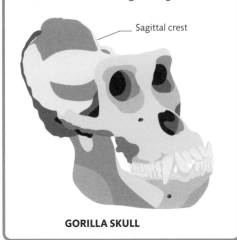

Sagittal crest

GORILLA SKULL

How the jaw works

The two temporomandibular joints between the lower jawbone and the skull each contain a disc of cartilage that provides a wider range of movement than is possible in other hinged joints, such as the elbow and knee. This disc is what allows the jaw to glide from side to side and move forwards and backwards when talking, chewing, or yawning.

WHAT CAUSES A CLICKING JAW?

If the protective disc of cartilage is displaced forwards, you may have a clicking jaw. The lower jawbone clicks against the zygomatic arch as you chew.

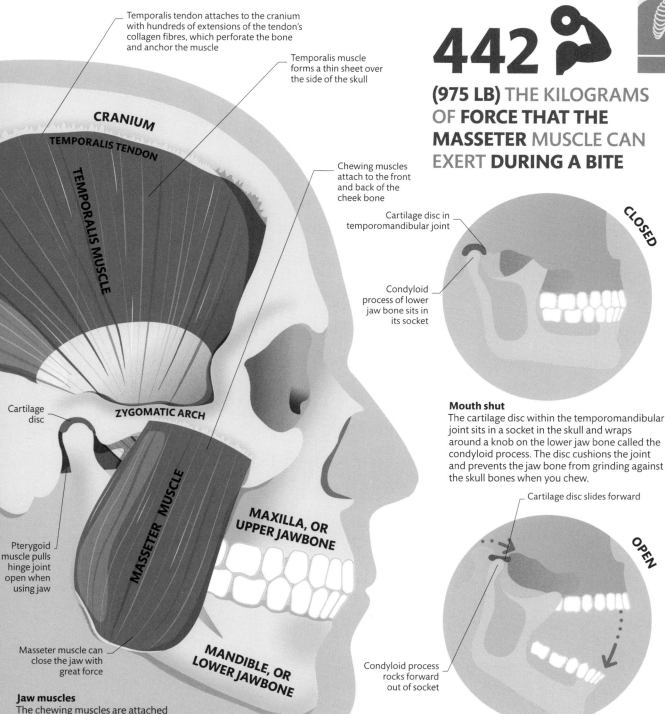

Temporalis tendon attaches to the cranium with hundreds of extensions of the tendon's collagen fibres, which perforate the bone and anchor the muscle

Temporalis muscle forms a thin sheet over the side of the skull

CRANIUM

TEMPORALIS TENDON

TEMPORALIS MUSCLE

Chewing muscles attach to the front and back of the cheek bone

442

(975 LB) THE KILOGRAMS OF FORCE THAT THE MASSETER MUSCLE CAN EXERT DURING A BITE

Cartilage disc in temporomandibular joint

Condyloid process of lower jaw bone sits in its socket

CLOSED

Cartilage disc

ZYGOMATIC ARCH

MASSETER MUSCLE

Pterygoid muscle pulls hinge joint open when using jaw

MAXILLA, OR UPPER JAWBONE

Mouth shut
The cartilage disc within the temporomandibular joint sits in a socket in the skull and wraps around a knob on the lower jaw bone called the condyloid process. The disc cushions the joint and prevents the jaw bone from grinding against the skull bones when you chew.

Cartilage disc slides forward

OPEN

Condyloid process rocks forward out of socket

Masseter muscle can close the jaw with great force

MANDIBLE, OR LOWER JAWBONE

Jaw muscles
The chewing muscles are attached to the skull. The strong temporalis and masseter muscles control the jaw as it grinds, snaps, and closes.

Mouth gaping
Both the lower jaw and the cushioning disc of cartilage can rock forward out of their socket, allowing your lower jawbone to hang open. Three fingers should fit between your upper and lower teeth.

Skin damage

Damaged skin, whether it is a superficial graze or a cut that penetrates deeper into the skin, lets infection enter the body. It is therefore important for healing to occur quickly, to prevent infections from spreading.

Wound healing

When the skin is breached, the first important step is to stem bleeding from a cut, or weeping fluid loss from a burn or blister. Some wounds need medical attention to seal them more firmly with stitches, sticky strips, or tissue glue. Covering the wound with a dressing will aid healing and reduce the chance of infection.

WHY DO SCABS ITCH?

During healing, when the cells move around the base of the wound, they begin to contract, helping to stitch the skin back together. As the tissues shrink, they stimulate specialized itch-sensitive nerve endings. Try not to scratch the scab off, though!

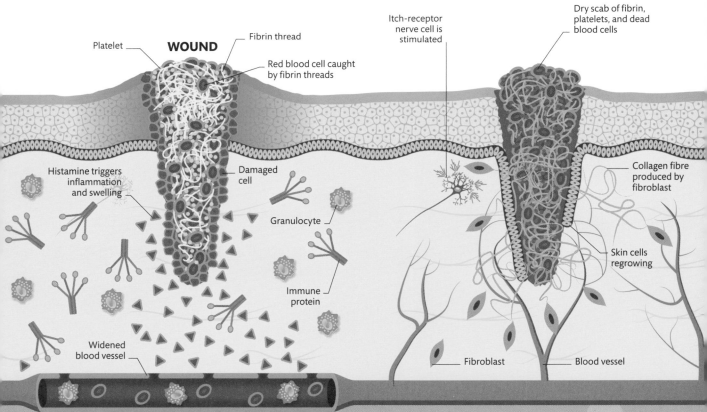

Platelet

WOUND

Fibrin thread

Red blood cell caught by fibrin threads

Itch-receptor nerve cell is stimulated

Dry scab of fibrin, platelets, and dead blood cells

Histamine triggers inflammation and swelling

Damaged cell

Collagen fibre produced by fibroblast

Granulocyte

Skin cells regrowing

Immune protein

Widened blood vessel

Fibroblast

Blood vessel

1 Clotting and inflammation
Platelets, which are fragments of blood cells, clump together to form a clot. Clotting factors form fibrin threads, which hold the clot in place. Inflammation floods the area with granulocytes and other cells and proteins of the immune system, which attack invading microbes.

2 Skin cells proliferate
Proteins called growth factors attract fibre-producing cells (fibroblasts), which move into the wound. They make granulation tissue, which is rich in tiny new blood vessels that grow into the area. Skin cells multiply to heal the wound from the base and sides.

WET AND DRY HEALING

When exposed to the air, a scab hardens so new skin cells have to burrow underneath and dissolve it away. Modern dressings help keep a wound moist so skin cells can leap-frog across the moist wound surface. This helps wounds heal more quickly, with less pain, less risk of infection, and less scarring.

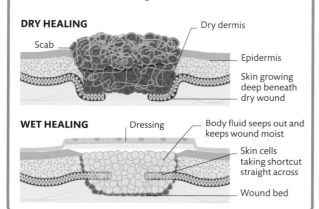

DRY HEALING

Scab — Dry dermis — Epidermis — Skin growing deep beneath dry wound

WET HEALING

Dressing — Body fluid seeps out and keeps wound moist — Skin cells taking shortcut straight across — Wound bed

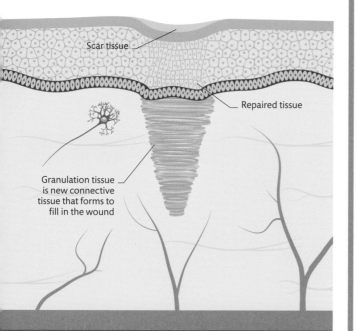

Scar tissue

Repaired tissue

Granulation tissue is new connective tissue that forms to fill in the wound

3 Remodelling
The surface skin cells have completed their job of growing over the damaged area and converting the scab into scar tissue. The scar shrinks to leave a red area that slowly becomes paler. Granulation tissue remains for a while.

Burns

If skin is heated above 49°C (120°F), its cells are damaged to cause a burn. Burns can also result from contact with chemicals and electricity.

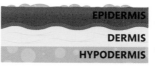

EPIDERMIS
DERMIS
HYPODERMIS

1st degree
Only the top layer of skin is injured, causing reddening and pain. Dead cells may peel after a few days.

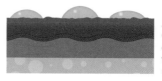

2nd degree
Cells in the deeper layers are destroyed and large blisters form. Enough live cells may remain to prevent scarring.

3rd degree
The full skin thickness is burned and skin grafts may be needed. There is a risk of scarring.

Blisters

A combination of heat, moisture, and friction may cause layers of skin to separate from each other and form a fluid-filled bubble, which protects the damaged skin. Covering them with a hydrocolloid gel blister plaster will soak up the fluid and form a cushioning, antiseptic environment so that the blister can heal faster.

Blister

Acne

Sebaceous glands release oil (sebum) onto the skin and hair. When the glands produce an excessive amount of sebum, the hair follicle can become clogged with sebum and dead skin cells to form a blackhead. Skin bacteria can infect the plug to cause a spot, which can leave a scar when it heals.

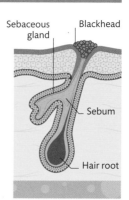

Sebaceous gland — Blackhead — Sebum — Hair root

Breaking and mending

A fracture is a break in a bone, which commonly results from an accident such as a fall, a road traffic collision, or a sports injury. Some fractures are relatively minor dents or hairline cracks that heal quickly, while severe impacts can shatter a bone into more than three pieces.

OPEN FRACTURE

Also known as a compound fracture, an open fracture is a nasty injury in which the skin is punctured either by the broken bone or by the impact that caused the injury. This means that infection can get in so antibiotics are usually given.

CLOSED FRACTURE

In a closed fracture, the skin remains intact. It is also known as a simple fracture. The injury is more likely to remain relatively sterile and avoid infection. Often, all that's needed is a cast to keep the bone still in the correct position for healing.

Immature bones are not fully mineralized and their bone may split on one side when bent, rather than breaking in two. This is known as a greenstick fracture and is often seen when a child falls out of a tree!

GREENSTICK FRACTURE

Spiral Fracture
A spiral fracture winds around the shaft of a long bone rather than breaking across. It results from a twisting force such as when a toddler lands on an outstretched leg when jumping.

A comminuted fracture occurs when a bone shatters into three or more pieces. This may need surgery to insert a plate and screws to hold loose bone fragments in position for stable healing.

A compression injury may cause the fractured ends of bone to collapse into one another and shorten the bone. The fracture must be stretched by traction – a gentle, steady action to pull the bones apart.

COMMINUTED FRACTURE

SPIRAL FRACTURE

COMPRESSION FRACTURE

Types of fractures

Bones may be broken by impacts and crushing, but also by repeated stress, such as marathon running. In young people, the most common broken bones are elbows and upper arms, which are often broken during play, or lower leg bones, often injured in sports and activities. Older people with brittle bones affected by osteoporosis (see p.50) are more likely to fracture hips and wrists.

THE BONE IN YOUR NOSE

Pinch your nose with your fingers and you will feel where the bone in the bridge of the nose is connected to cartilage at the tip. When you break your nose, it's the bone at the top that gets fractured.

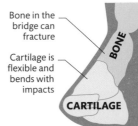

Bone in the bridge can fracture

Cartilage is flexible and bends with impacts

BONE

CARTILAGE

Dislocation

If the ligaments supporting a mobile joint are stretched during a wrenching accident, the bones can slip out of place, causing a joint dislocation. It is most common in the shoulder, finger, and thumb joints. To treat a dislocation, medics fit the bones back into place and keep the joint still with a cast or a sling, so that the ligaments can heal. Some joints, such as the shoulder, can dislocate again and again if the ligaments remain slack.

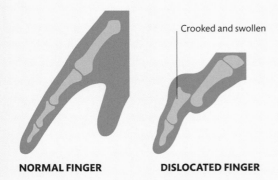

Crooked and swollen

NORMAL FINGER   **DISLOCATED FINGER**

Dislocated joint
The finger joints may dislocate if you catch a ball awkwardly. It causes pain, swelling, and an obviously abnormal shape. Once the dislocated bones are repositioned (after an X-ray to rule out a fracture) the fingers are splinted together to allow healing.

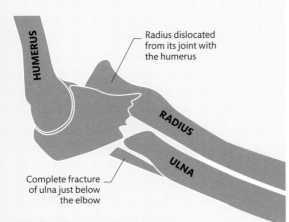

HUMERUS

Radius dislocated from its joint with the humerus

RADIUS

ULNA

Complete fracture of ulna just below the elbow

Break and dislocation together
When a fracture is close to a joint, the ligaments may give way so both a fracture and a dislocation occur. This is commonly seen at the elbow when the ulna fractures, and the head of the radius is displaced.

Healing

Bones can heal like any other living tissue, but the process takes longer as minerals must be laid down until the bone is strong again. A broken bone is immobilized by a rigid cast around the body part. If it needs firmer support, surgical screws or a metal plate may be inserted. The fracture then heals in several stages.

1 Immediate response
The fracture site quickly fills with blood to form a massive clot. The tissue around the injury forms a bruise-like swelling. The area is painful, inflamed, and some bone cells die due to poor circulation.

Ruptured blood vessel

Periosteum (the bone's "skin") is broken

Blood-filled swelling

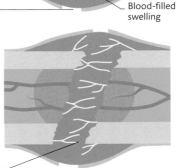

2 Three days later
Blood capillaries grow into the blood clot and the damaged tissue is slowly broken down, absorbed, and removed by scavenger cells. Specialized cells move into the area and start laying down collagen fibres that act like scaffolding for bone cells.

Collagen fibres

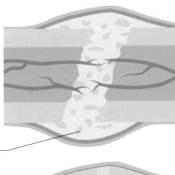

3 Three weeks later
Collagen fibres join up across the fracture to link the bone ends. The repair process forms a swelling, called a callus, which is initially formed of cartilage. This provides weak support which can easily re-fracture if moved too early.

Callus

4 Three months later
Cartilage within the repair tissue is replaced with strong spongy bone and compact bone forms around the outer edge of the fracture. Remodelling of the healing fracture will remove the excess callus and the bone swelling eventually straightens out.

Healed fracture

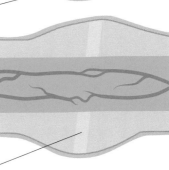

Wearing thin

Cells in our bones are constantly remodelling our skeletons by dissolving old bone and laying down new bone. However, sometimes this process becomes unbalanced, leading to a variety of problems, not all of which are easily solved.

When bones wear out

The brittle bone disease, osteoporosis, develops when not enough new bone is made to replace the old. This imbalance can happen if you don't eat enough calcium-rich foods or if you don't top up your vitamin D – either in your diet or by getting enough sun (see p.33) – which the body needs to absorb calcium efficiently. It can also result from hormone changes in later life, such as when female oestrogen levels fall after the menopause. Osteoporosis produces few symptoms, but the first indication is often when a fracture of the hip or wrist occurs after a minor fall.

Depleted outer layer of compact bone

Strong outer layer of compact bone

OSTEOPOROTIC BONE

Spongy interior

Brittle interior of weakened bone

HEALTHY BONE

BONE EXERCISE

Regular exercise stimulates the production of new bone tissue. High-impact exercises, such as aerobics, jogging, or racquet sports are best, but any weight-bearing exercise, including gentle yoga or tai chi, helps to stimulate strengthening at areas where bone is stressed.

In this yoga exercise, the shin bone is under stress

Healthy bone
Healthy bone has a strong, thick, outer layer of dense, compact tissue and a good, underlying network of spongy bone. This structure shows up clearly on X-rays and is strong enough to withstand minor blows, such as a fall onto your outstretched hands.

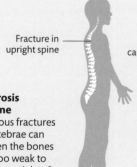

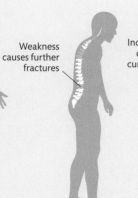

Fracture in upright spine

Weakness causes further fractures

Increasing damage curves the spine

Osteoporosis in the spine
Spontaneous fractures of the vertebrae can occur when the bones become too weak to support the weight of the upper body. This causes pain and leads to an increasingly curved spine.

EARLY STAGE

LATER STAGE

ADVANCED STAGE

HOW COMMON IS OSTEOPOROSIS ?

Worldwide, one in three women and one in five men over the age of 50 experience an osteoporotic bone fracture. Smoking, alcohol, and lack of exercise increase the risk of injury.

MILK

PEACHES

BONE

BROCCOLI

CHEESE

Topping up your calcium
A balanced diet containing good supplies of calcium-rich foods is essential at all stages of life to help prevent osteoporosis. Good dietary sources of calcium include dairy products, some fruit and vegetables, nuts, seeds, pulses, eggs, canned fish (with bones), and fortified bread.

ORANGES

FISH

SOYA BEANS

Osteoporotic bone
Brittle bones have only a thin outer layer of dense, compact bone and fewer struts within the underlying network of spongy bone. Thin bones barely show on X-rays and may fracture in a simple fall.

When joints becomes weak

Joints are subject to a lot of wear and tear, which leads to a type of inflammation called osteoarthritis. This is especially common in weight-bearing joints, such as the knee and the hip, causing increasing pain, stiffness, and restricted movements. The joint cartilage weakens and flakes away, leaving the bone ends to rub together and form bony outgrowths.

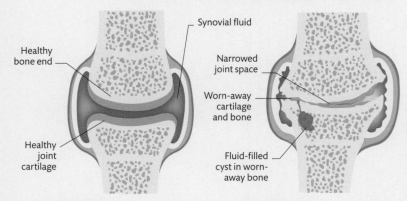

Synovial fluid

Healthy bone end

Narrowed joint space

Worn-away cartilage and bone

Healthy joint cartilage

Fluid-filled cyst in worn-away bone

Healthy joint
In a healthy joint, the two bones are cushioned with cartilage and are separated by a film of lubricant called synovial fluid.

Arthritic joint
In an arthritic joint, the joint cartilages are eroding. The bones grind together and the synovial fluid is unable to lubricate the joint.

JOINT REPLACEMENT

Osteoarthritis is treated simply with painkillers, but when symptoms interfere with a person's quality of life, a better solution is to replace the worn-out joint with an artificial one made of metal, plastic, or ceramic. However, even artificial joints eventually wear out, and may need replacing every 10 years or so. A commonly replaced joint is the hip joint.

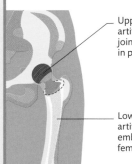

Upper part of artificial hip joint, embedded in pelvis

Lower part of artificial hip joint, embedded in femur (thigh bone)

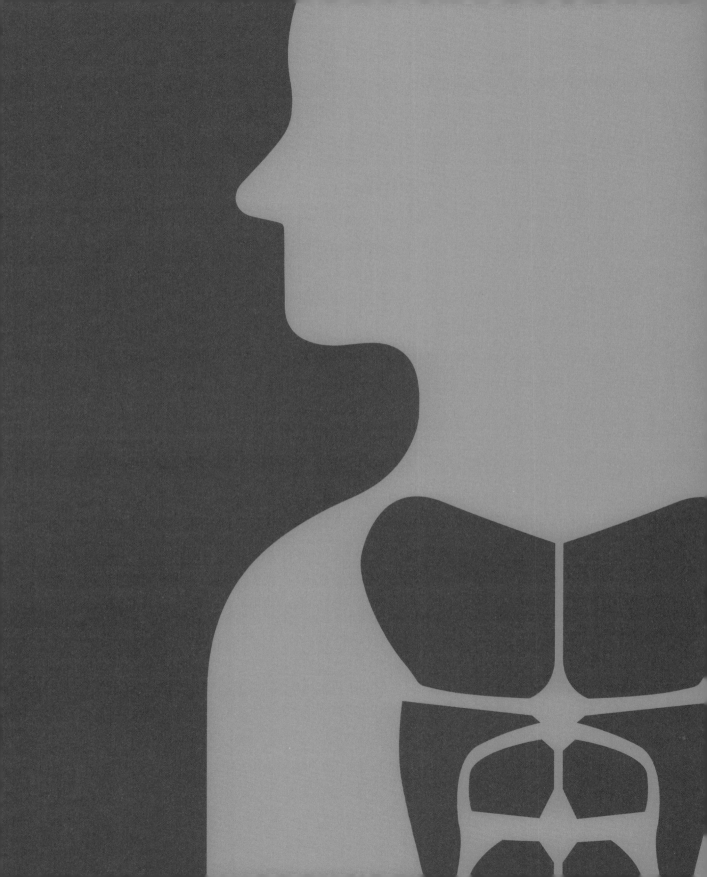

ON THE
MOVE

Pulling power

Muscles carry out all the body's movements and are attached to bones by tendons. The tendons are made of strong, connective tissue that can stretch to help deal with the forces produced during movement.

Teamwork

Muscles can only pull, they cannot push. They therefore work in pairs or teams that work in opposition to each other. When one set of muscles contracts, the other relaxes to bend a joint. They swap roles to straighten the joint again. For example, contraction of the biceps bends the elbow, while contraction of the triceps straightens it as the biceps relaxes. Muscle can only "push" indirectly, via levers.

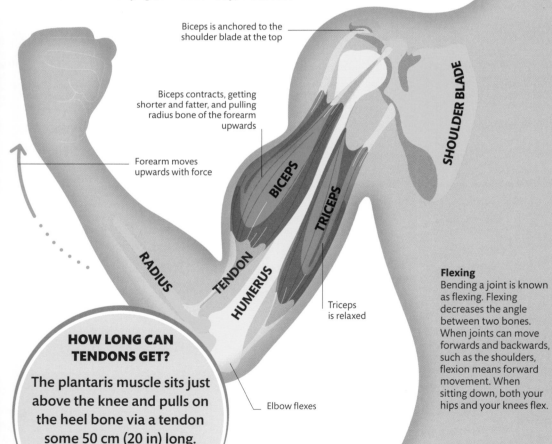

Biceps is anchored to the shoulder blade at the top

Biceps contracts, getting shorter and fatter, and pulling radius bone of the forearm upwards

Forearm moves upwards with force

SHOULDER BLADE

BICEPS

TRICEPS

RADIUS

TENDON

HUMERUS

Triceps is relaxed

Elbow flexes

HOW LONG CAN TENDONS GET?

The plantaris muscle sits just above the knee and pulls on the heel bone via a tendon some 50 cm (20 in) long. The Achilles is the strongest and thickest tendon.

Flexing
Bending a joint is known as flexing. Flexing decreases the angle between two bones. When joints can move forwards and backwards, such as the shoulders, flexion means forward movement. When sitting down, both your hips and your knees flex.

Extending
Extension is the opposite of flexion and increases the angle between two bones. When joints can move forwards and backwards, such as the hips, extension means backwards movement. When standing, both your hips and your knees extend.

Body levers

A lever allows movement to occur around a point called a fulcrum. A first class lever has the fulcrum in the middle. A second class lever places the load between the effort and fulcrum. In a third class lever, the effort occurs between the load and fulcrum – like using a pair of tweezers.

KEY:

▲ Fulcrum ↑ Direction of force ↑ Movement of load

First-class lever

Neck muscles work like first class levers. When they contract, they force your chin up on the opposite side of the fulcrum (a joint between your skull and spine).

NECK MUSCLE

CALF MUSCLE

Body rises a little way, but with great force

Second-class lever

The calf muscle can act as a second-class lever by pulling when the foot is on the ground. The foot then bends at the base of the toe so the entire weight of the body is raised on tiptoe.

Third-class lever

The biceps acts as a third-class lever. Pulling close to the fulcrum – the elbow – it moves the bones only a little, but creates a lot of movement for the hand at the end of the lever. A small effort translates into a big movement.

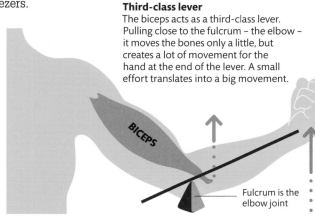

BICEPS

Fulcrum is the elbow joint

THE ACHILLES TENDON IS STRONG ENOUGH TO SUPPORT MORE THAN 10 TIMES YOUR BODY WEIGHT WHEN RUNNING.

Triceps is anchored to the shoulder blade and humerus at the top end

Biceps is relaxed, and can lengthen, allowing triceps to extend the elbow

Forearm moves down

Muscle's tendon splits to pull on four fingertips

Finger extender muscle is anchored to the upper arm bone at one end

BICEPS

TRICEPS

ULNA

Elbow extends (straightens)

Triceps contracts, pulling on the ulna bone of the forearm

Remote control

Muscles pull on bones via tendons. However, the tendons can be very long, and the muscles far from the joints they are operating. Amazingly, there are no muscles at all in the fingers. All of their movement is made by remote control – by muscles in the hand and arm.

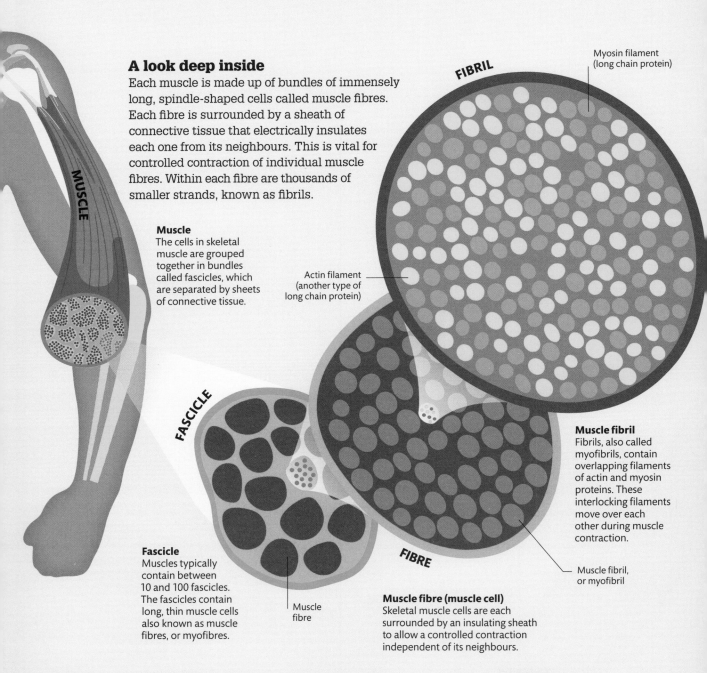

A look deep inside

Each muscle is made up of bundles of immensely long, spindle-shaped cells called muscle fibres. Each fibre is surrounded by a sheath of connective tissue that electrically insulates each one from its neighbours. This is vital for controlled contraction of individual muscle fibres. Within each fibre are thousands of smaller strands, known as fibrils.

FIBRIL

Myosin filament
(long chain protein)

Muscle
The cells in skeletal muscle are grouped together in bundles called fascicles, which are separated by sheets of connective tissue.

Actin filament
(another type of
long chain protein)

MUSCLE

Muscle fibril
Fibrils, also called myofibrils, contain overlapping filaments of actin and myosin proteins. These interlocking filaments move over each other during muscle contraction.

Muscle fibril,
or myofibril

FASCICLE

FIBRE

Fascicle
Muscles typically contain between 10 and 100 fascicles. The fascicles contain long, thin muscle cells also known as muscle fibres, or myofibres.

Muscle
fibre

Muscle fibre (muscle cell)
Skeletal muscle cells are each surrounded by an insulating sheath to allow a controlled contraction independent of its neighbours.

How do muscles pull?

Muscle cells carry out all body movements. Some muscles are under voluntary control and only contract when you want them to. Others contract automatically to keep your body working smoothly. Muscle cells are able to contract due to actin and myosin molecules.

Miracle molecules

Actin and myosin filaments are arranged in units called sarcomeres. When a muscle receives a signal to contract, the myosin filaments repeatedly pull the actin filaments along so that they slide closer and closer together. This makes the muscle shorten. They slide apart when the muscle relaxes again.

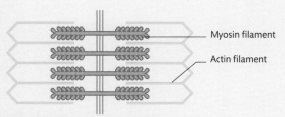

Myosin filament

Actin filament

SARCOMERE OF RELAXED MUSCLE

1 Myosin energized
Myosin head is energized by ATP molecule (produced from sugars and oxygen).

Actin

Myosin head

2 Myosin head sticks to actin
The energized myosin head sticks to the actin filament, forming a cross bridge.

Head sticks to actin

Myosin energized

3 Head pivots
The myosin head releases energy and pivots, sliding the actin filament over. The cross bridge weakens.

Actin pulled along

Head pivots

4 Re-energizing
The cross bridge releases and the myosin head is re-energized. These steps happen many times during a single contraction.

Head detaches

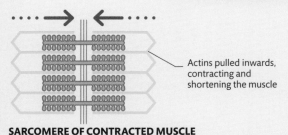

Actins pulled inwards, contracting and shortening the muscle

SARCOMERE OF CONTRACTED MUSCLE

FAST AND SLOW TWITCHING

Muscles have two types of fibre. Fast-twitch fibres reach peak contraction – the peak of their power output – in 50 milliseconds, but fatigue after a few minutes. Slow-twitch fibres take 110 milliseconds to reach peak contraction but do not tire. The explosive power needed by sprinters means that they tend to have more fast-twitch fibres. Long-distance runners usually have more slow-twitch fibres that don't fatigue as quickly as fast-twitch fibres.

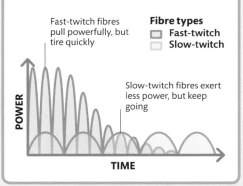

Fast-twitch fibres pull powerfully, but tire quickly

Fibre types
- Fast-twitch
- Slow-twitch

Slow-twitch fibres exert less power, but keep going

POWER

TIME

CRAMP

Sometimes voluntary muscle may contract involuntarily, causing painful cramping. This occurs when chemical imbalances – for example, when poor circulation leads to low oxygen levels and a build-up of lactic acid – interfere with the release of the cross bridges. Gentle stretching and rubbing the contracted muscle stimulates circulation and helps muscle relaxation.

FAST-TWITCH FIBRES CAN CONTRACT AT A RATE OF **30–50 TIMES** PER SECOND.

Working, stretching, pulling, braking

Muscles shorten and pull on bones to bend joints and create movement. However, they also contract without any movement to create power and tension, which can hold a weight steady. If the weight is too great to hold, muscles can even contract and lengthen as they brake the movement of the weight.

Pulling and shortening

If you contract your biceps muscle when lifting a gym weight during a "biceps curl", the muscle shortens, producing a movement in the direction of the contraction. The force generated by the muscle is greater than the weight or force it pulls against. The muscles contain both contractile fibres, which shorten, and elastic fibres, which stretch if tension increases. During a shortening contraction, the contractile fibres cause the muscle length to change, but the tension in the elastic fibres remain unchanged.

WHY WARM UP BEFORE EXERCISE?

Doing exercise to loosen muscles and increase blood flow helps to limit muscle injuries, such as tears and strains, which can occur with sudden vigorous movement.

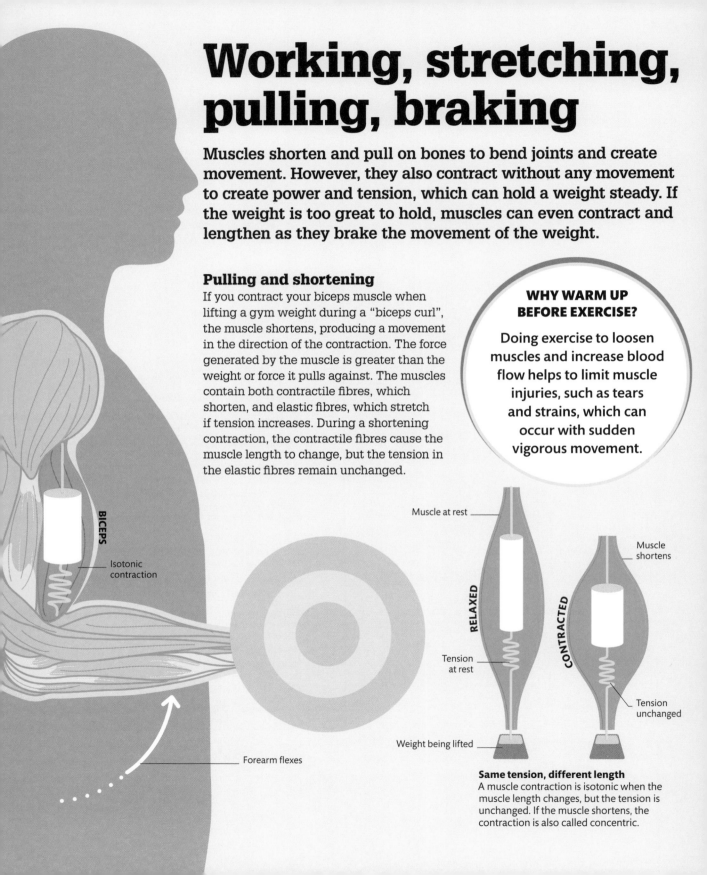

BICEPS

Isotonic contraction

Forearm flexes

Muscle at rest

RELAXED

Tension at rest

Weight being lifted

Muscle shortens

CONTRACTED

Tension unchanged

Same tension, different length
A muscle contraction is isotonic when the muscle length changes, but the tension is unchanged. If the muscle shortens, the contraction is also called concentric.

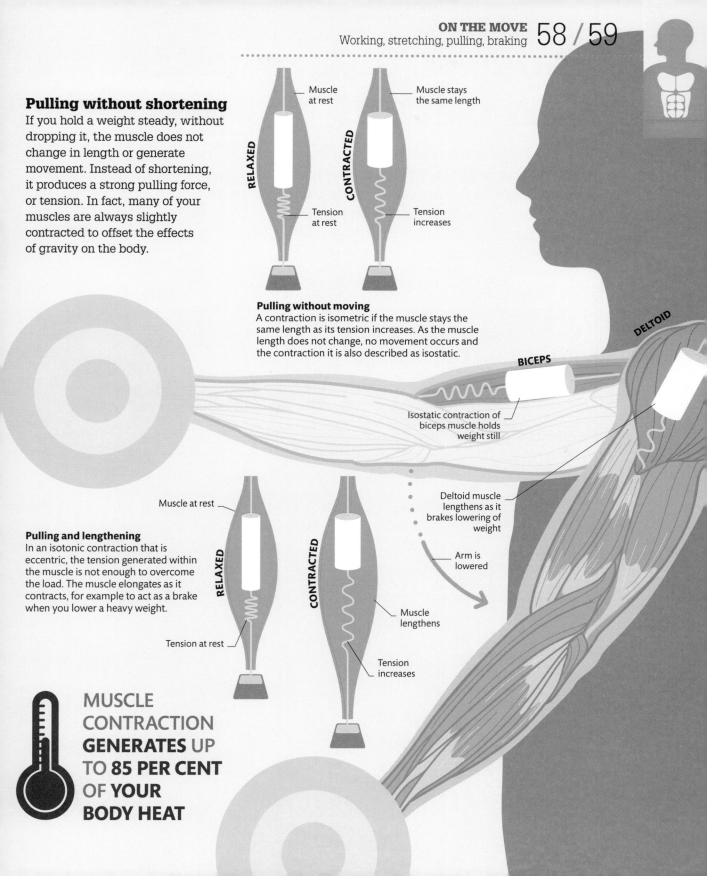

Pulling without shortening

If you hold a weight steady, without dropping it, the muscle does not change in length or generate movement. Instead of shortening, it produces a strong pulling force, or tension. In fact, many of your muscles are always slightly contracted to offset the effects of gravity on the body.

RELAXED

Muscle at rest

Tension at rest

CONTRACTED

Muscle stays the same length

Tension increases

Pulling without moving

A contraction is isometric if the muscle stays the same length as its tension increases. As the muscle length does not change, no movement occurs and the contraction it is also described as isostatic.

DELTOID

BICEPS

Isostatic contraction of biceps muscle holds weight still

Pulling and lengthening

In an isotonic contraction that is eccentric, the tension generated within the muscle is not enough to overcome the load. The muscle elongates as it contracts, for example to act as a brake when you lower a heavy weight.

RELAXED

Muscle at rest

Tension at rest

CONTRACTED

Muscle lengthens

Tension increases

Deltoid muscle lengthens as it brakes lowering of weight

Arm is lowered

MUSCLE CONTRACTION **GENERATES** UP TO **85 PER CENT** OF **YOUR BODY HEAT**

Sensory input, action output

The brain and spinal cord form the central nervous system. They receive sensory input from all over the body via a vast network of "sensory" nerve cells. In response to the sensory information, the brain and spinal cord send instructions down "motor" nerve cells to control your actions.

IT CAN TAKE YOUR BRAIN UP TO **400 MILLISECONDS** TO **PROCESS INCOMING INFORMATION** BEFORE YOU **BECOME CONSCIOUS** OF IT

HOW FAST?

Reflex reactions are much faster than reaction times routed via the brain. This is true of reactions to visual, hearing, or touch sensations.

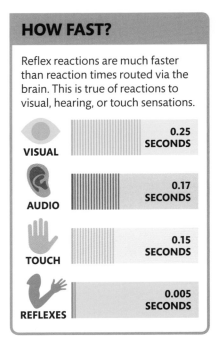

VISUAL	0.25 SECONDS
AUDIO	0.17 SECONDS
TOUCH	0.15 SECONDS
REFLEXES	0.005 SECONDS

INPUT (SENSORY NERVES)

Consulting the brain

If a movement requires conscious thought, such as listening for a starter gun, the sensory signal travels up the spinal cord to the brain for processing before the body takes action. Some conscious actions become relatively automatic and are performed on "autopilot", without thinking. In fact, most nerve signals sent to and from the brain, just to keep the body functioning properly, occur subconsciously.

Sprinter in position

Ear interprets gunshot as an audio signal

Expecting the signal
A sprinter is poised at the start line, waiting for the gunshot to start running.

Audio cue
The starter gun sounds. Audio waves reach the ear, which sends sensory messages to the brain.

Taking the brain out of the loop

Survival sometimes requires instant responses that bypass the brain and happen as automatic reflexes. Reflex pathways are routed via the spinal cord to avoid the delays that would occur if the messages travelled via the brain. When a reflex action is performed, the brain may be informed straight afterwards.

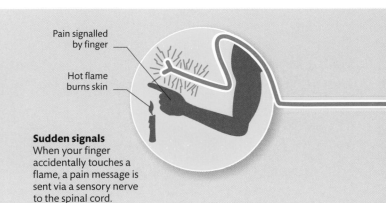

Pain signalled by finger

Hot flame burns skin

Sudden signals
When your finger accidentally touches a flame, a pain message is sent via a sensory nerve to the spinal cord.

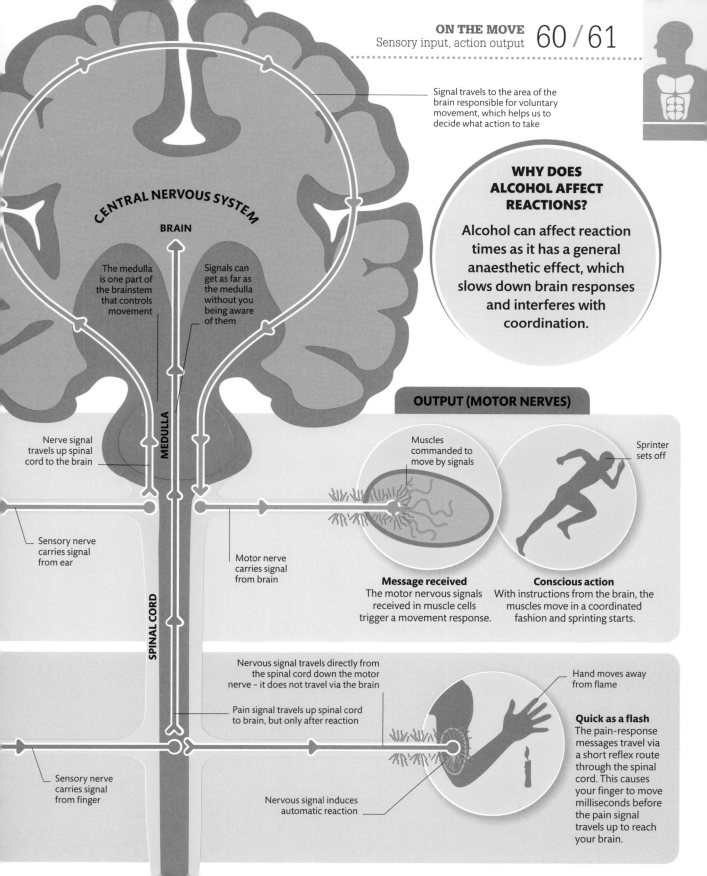

Signal travels to the area of the brain responsible for voluntary movement, which helps us to decide what action to take

CENTRAL NERVOUS SYSTEM

BRAIN

The medulla is one part of the brainstem that controls movement

Signals can get as far as the medulla without you being aware of them

MEDULLA

WHY DOES ALCOHOL AFFECT REACTIONS?

Alcohol can affect reaction times as it has a general anaesthetic effect, which slows down brain responses and interferes with coordination.

Nerve signal travels up spinal cord to the brain

Sensory nerve carries signal from ear

SPINAL CORD

Motor nerve carries signal from brain

OUTPUT (MOTOR NERVES)

Muscles commanded to move by signals

Sprinter sets off

Message received
The motor nervous signals received in muscle cells trigger a movement response.

Conscious action
With instructions from the brain, the muscles move in a coordinated fashion and sprinting starts.

Nervous signal travels directly from the spinal cord down the motor nerve – it does not travel via the brain

Pain signal travels up spinal cord to brain, but only after reaction

Hand moves away from flame

Quick as a flash
The pain-response messages travel via a short reflex route through the spinal cord. This causes your finger to move milliseconds before the pain signal travels up to reach your brain.

Sensory nerve carries signal from finger

Nervous signal induces automatic reaction

The control centre

The brain coordinates all body functions. It contains billions of nerve cells whose interconnections make it the most complex of all your organs. The brain can process thoughts, actions, and emotions simultaneously. Despite popular belief, you use all of your brain although the exact function of some areas remains elusive.

Inside the brain

The brain is divided into two main parts – the higher brain and the primitive brain. The higher brain is the larger of the two and consists of the cerebrum, which is divided into two halves called the left and right hemispheres. The higher brain is where conscious thoughts are processed. The more primitive part of the brain, which connects with the spinal cord, is where your body's automatic functions, such as breathing and blood pressure, are controlled.

Grey matter
The darker outer layer of the brain is composed mainly of nerve cell bodies, some of which cluster together to form nerve ganglia.

NERVE CELL BODY

White matter
The fine nerve filaments, or axons, which carry electrical impulses away from each nerve cell, form the paler tissue beneath the grey matter.

Axon   Nerve

NERVE

GREY MATTER

Primitive brain
The cerebellum, thalamus, and brainstem deal with instinctive responses and automatic functions, such as body temperature and sleep-wake cycles. This part of the brain also generates primitive emotions, such as anger and fear. The cerebellum coordinates muscle movements and balance.

The brain at work

When you learn a skill, new connections form between the brain cells that are used. This means that unfamiliar actions start to become automatic. The amount of practice a golfer does is reflected in the active areas of their brain when they swing their club.

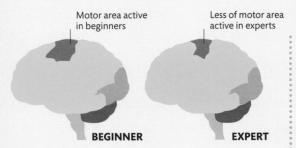

Motor area active in beginners

Less of motor area active in experts

BEGINNER   **EXPERT**

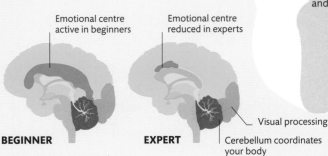

Emotional centre active in beginners

Emotional centre reduced in experts

BEGINNER   **EXPERT**

Visual processing

Cerebellum coordinates your body

Outer cerebral activity
As you practise your shots, less of your motor area will be stimulated as the once unfamiliar action becomes more refined. Areas devoted to coordination and visual processing in both beginners and experts remain the same.

Inner cerebral activity
A cross-section of the brain reveals that the brain's emotional centre is active in beginners, who may deal with anxiety or embarrassment. Expert golfers learn to control their emotions and concentrate solely on taking the shot.

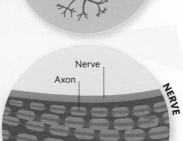

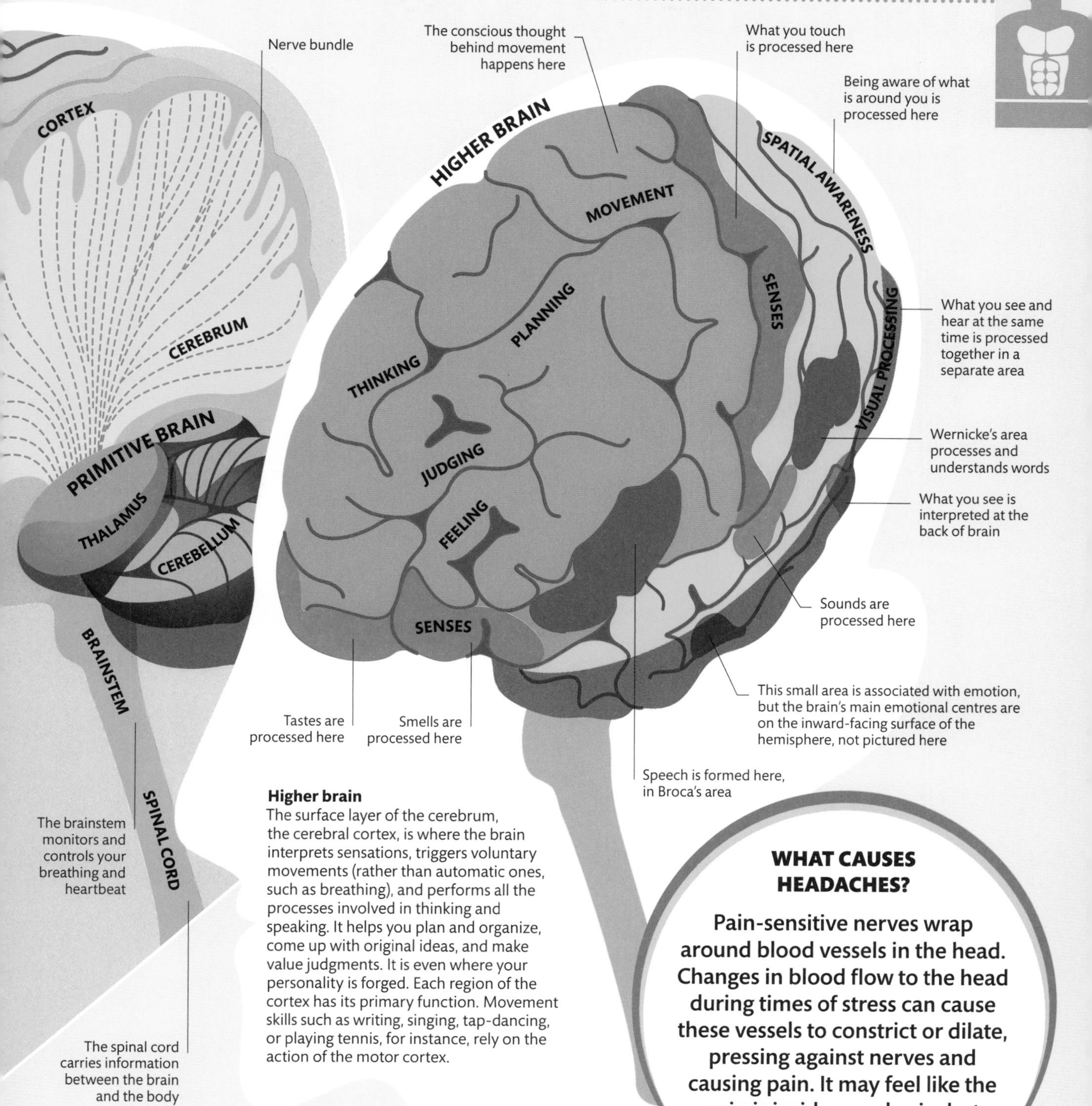

CORTEX

Nerve bundle

The conscious thought
behind movement
happens here

What you touch
is processed here

Being aware of what
is around you is
processed here

HIGHER BRAIN

MOVEMENT

SPATIAL AWARENESS

CEREBRUM

PLANNING

SENSES

THINKING

VISUAL PROCESSING

What you see and
hear at the same
time is processed
together in a
separate area

PRIMITIVE BRAIN

JUDGING

THALAMUS

FEELING

Wernicke's area
processes and
understands words

CEREBELLUM

What you see is
interpreted at the
back of brain

SENSES

Sounds are
processed here

BRAINSTEM

Tastes are
processed here

Smells are
processed here

SPINAL CORD

This small area is associated with emotion,
but the brain's main emotional centres are
on the inward-facing surface of the
hemisphere, not pictured here

The brainstem
monitors and
controls your
breathing and
heartbeat

Speech is formed here,
in Broca's area

Higher brain
The surface layer of the cerebrum,
the cerebral cortex, is where the brain
interprets sensations, triggers voluntary
movements (rather than automatic ones,
such as breathing), and performs all the
processes involved in thinking and
speaking. It helps you plan and organize,
come up with original ideas, and make
value judgments. It is even where your
personality is forged. Each region of the
cortex has its primary function. Movement
skills such as writing, singing, tap-dancing,
or playing tennis, for instance, rely on the
action of the motor cortex.

The spinal cord
carries information
between the brain
and the body

WHAT CAUSES
HEADACHES?

**Pain-sensitive nerves wrap
around blood vessels in the head.
Changes in blood flow to the head
during times of stress can cause
these vessels to constrict or dilate,
pressing against nerves and
causing pain. It may feel like the
pain is inside your brain, but
no pain-sensitive nerves
are there!**

Communication hub

When you think or act, it is not a single region of the brain that becomes active, but rather a network of cells spread across several brain regions. It is these patterns of activity that command your mind and body.

BRAIN

CORPUS CALLOSUM

Brain hemispheres

Your brain is divided into two hemispheres. Structurally, they are almost identical, however each of them is responsible for certain tasks. The left hemisphere controls the right side of the body and (in most people) is responsible for language and speech. The right hemisphere controls the left side of the body and is responsible for an awareness of your surroundings, sensory information, and creativity. The two halves of your brain work together, communicating through a nerve superhighway called the corpus callosum.

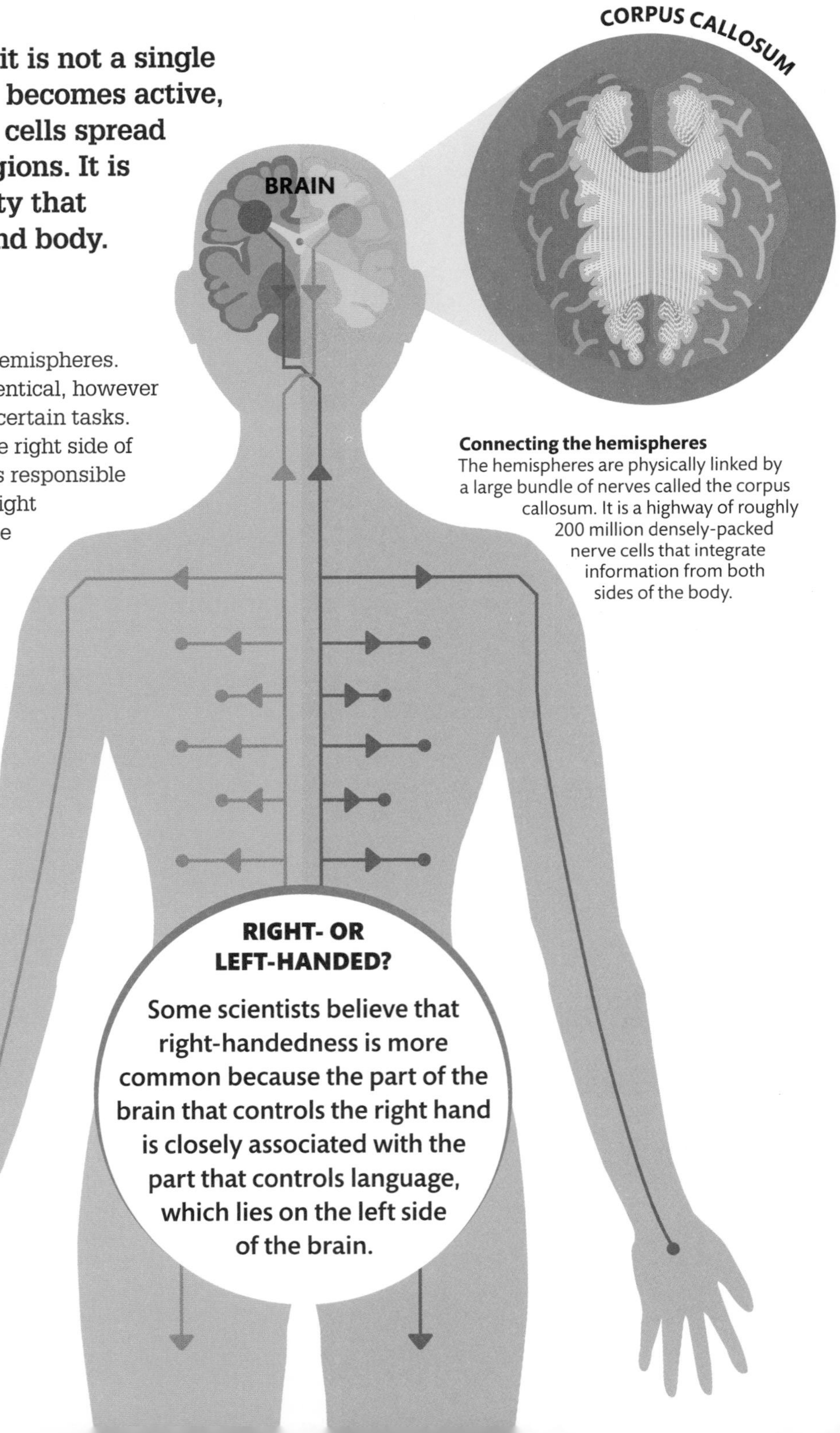

Connecting the hemispheres
The hemispheres are physically linked by a large bundle of nerves called the corpus callosum. It is a highway of roughly 200 million densely-packed nerve cells that integrate information from both sides of the body.

Controling opposite sides
Each side of your body sends information to, and is controlled by, the opposite hemisphere of the brain. Information travels between them by a nerve network that spreads to every inch of your body.

RIGHT- OR LEFT-HANDED?

Some scientists believe that right-handedness is more common because the part of the brain that controls the right hand is closely associated with the part that controls language, which lies on the left side of the brain.

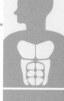

THE **BRAIN** CONTAINS **86 BILLION NERVE CELLS** JOINED BY **100 TRILLION CONNECTIONS** – MORE THAN THE NUMBER OF STARS IN THE MILKY WAY

Nerve pathway that connects brain regions

One of many nodes in brain active while playing chess

Networks in the brain

To perform the simplest action, such as walking, or a complex manoeuvre, such as a dance, you rarely use just one area of your brain. In fact, networks of connected areas all over the brain are often activated as you go about your day. By looking for regions consistently activated together, researchers can track the flow of information around the brain. These networks can change during your lifetime as you learn new skills and information, and as a result new nerve pathways are made. Unused nerve pathways may be pruned as you grow older.

Multiple areas at work
When you play chess, you use many regions of your brain. Not only do you use your visual processing region, you may also activate your memory and planning areas to recall previous games and establish a strategy.

This nerve cell is connected to four others, forming a network across the brain

Physical connections
Scientists can trace the physical connections between nerve cells in the brain. The density of nerve pathways indicates which brain regions communicate the most.

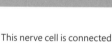

Nerve activity is shown as areas that light up on some brain scans

Active brain areas
The electrical activity that nerve cells generate can be picked up on certain types of brain scans. Looking at these scans can shed light on which brain regions are most active during particular tasks.

DEFAULT MODE

When you are relaxed and not focusing on the world around you, your brain shows a specific pattern of activity; this is called the default mode network. It is thought that this network helps to generate thoughts as your mind wanders, and may be linked with creativity, self-reflection, and moral reasoning.

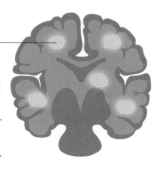

CREATIVE THOUGHTS

DAYDREAMER

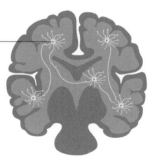

Sparking into life

Nerves transmit electrical messages around the body in milliseconds. Each nerve is like a cable of insulated wires, and each wire is called a nerve fibre, or axon. An axon is the main part of a single, immensely long cell – called a neuron – whose job it is to pass on the signal.

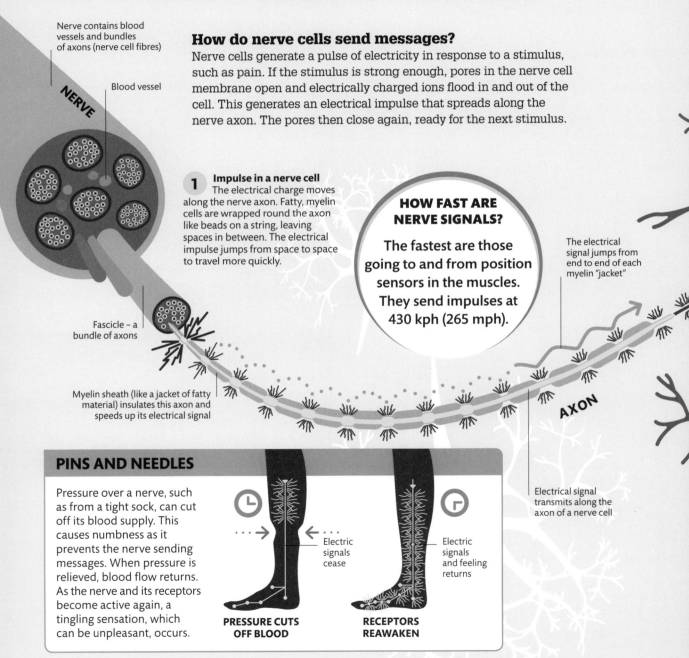

Nerve contains blood vessels and bundles of axons (nerve cell fibres)

Blood vessel

NERVE

How do nerve cells send messages?

Nerve cells generate a pulse of electricity in response to a stimulus, such as pain. If the stimulus is strong enough, pores in the nerve cell membrane open and electrically charged ions flood in and out of the cell. This generates an electrical impulse that spreads along the nerve axon. The pores then close again, ready for the next stimulus.

1 Impulse in a nerve cell
The electrical charge moves along the nerve axon. Fatty, myelin cells are wrapped round the axon like beads on a string, leaving spaces in between. The electrical impulse jumps from space to space to travel more quickly.

HOW FAST ARE NERVE SIGNALS?

The fastest are those going to and from position sensors in the muscles. They send impulses at 430 kph (265 mph).

The electrical signal jumps from end to end of each myelin "jacket"

Fascicle – a bundle of axons

Myelin sheath (like a jacket of fatty material) insulates this axon and speeds up its electrical signal

AXON

Electrical signal transmits along the axon of a nerve cell

PINS AND NEEDLES

Pressure over a nerve, such as from a tight sock, can cut off its blood supply. This causes numbness as it prevents the nerve sending messages. When pressure is relieved, blood flow returns. As the nerve and its receptors become active again, a tingling sensation, which can be unpleasant, occurs.

Electric signals cease

Electric signals and feeling returns

PRESSURE CUTS OFF BLOOD

RECEPTORS REAWAKEN

Dendrites connect to other nerve cells

THE GAP BETWEEN NERVE CELLS IS LESS THAN **1 TRILLIONTH** THE WIDTH OF A **HUMAN HAIR**

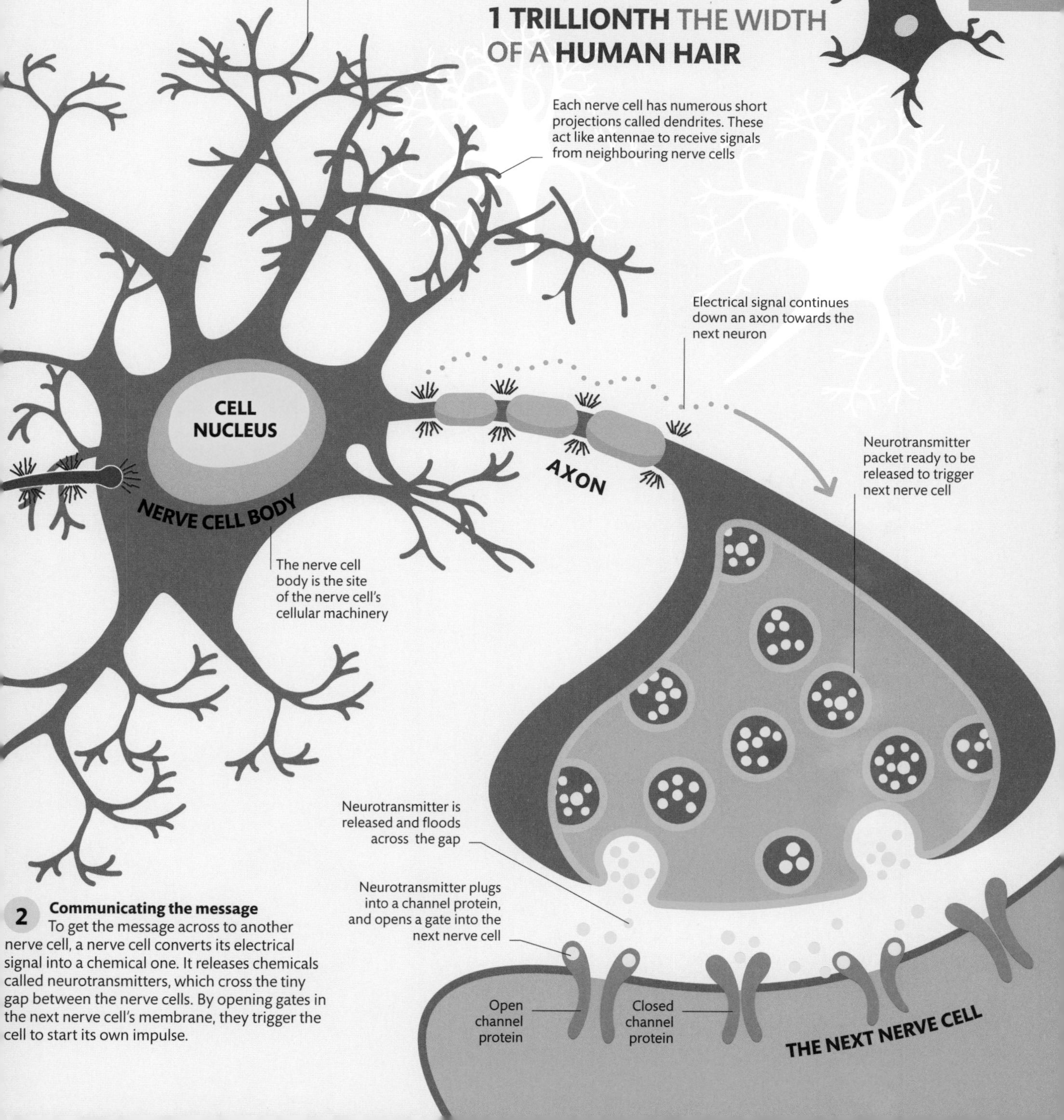

Each nerve cell has numerous short projections called dendrites. These act like antennae to receive signals from neighbouring nerve cells

Electrical signal continues down an axon towards the next neuron

CELL NUCLEUS

AXON

Neurotransmitter packet ready to be released to trigger next nerve cell

NERVE CELL BODY

The nerve cell body is the site of the nerve cell's cellular machinery

Neurotransmitter is released and floods across the gap

Neurotransmitter plugs into a channel protein, and opens a gate into the next nerve cell

2 **Communicating the message**
To get the message across to another nerve cell, a nerve cell converts its electrical signal into a chemical one. It releases chemicals called neurotransmitters, which cross the tiny gap between the nerve cells. By opening gates in the next nerve cell's membrane, they trigger the cell to start its own impulse.

Open channel protein

Closed channel protein

THE NEXT NERVE CELL

RELAXATION

BRAIN

BRAINSTEM

Pupils constrict
Normal pupil responses return, so that they constrict, or narrow, in bright light and dilate only when light levels are reduced.

Small airways narrow
When relaxed, the airways within your lungs return to their normal size, allowing for a regular intake of oxygen.

Blood vessels narrow
Your arteries return to their normal size when you are relaxed. Blood flow is evenly distributed across the body.

Heart rate decreases
Your heart rate returns to your normal resting rate as you relax. However, resting heart rate can vary with your fitness level.

Liver stores sugars
When you are relaxed, your liver saves up energy. Any excess sugar you ingest is packed away, or converted into fat and then stored as extra tissue.

SPINAL CORD

ACTION

Pupils dilate
Dilation, or widening, of your pupils allows more light to enter your eyes. This occurs in darkness to improve your vision.

Small airways expand
The bronchioles, tiny airways in your lungs, widen to allow more air in. You absorb more oxygen, which muscles use for fuel, if a quick getaway is needed.

Widening arteries
Arteries to your muscles and brain dilate to provide these organs with more oxygen, so you act faster and think more quickly. As a result, blood is diverted away from your skin, making you pale.

Heart rate increases
Your pulse rises to 100 beats per minute or more so that more blood is sent to the lungs to collect oxygen and to the body to distribute oxygen.

Liver releases sugars
Your liver acts as your body's engine. It converts glucose, a sugar, to energy, using stores in your body. Your muscles need energy in order to move.

Digestion stimulated
In the absence of stress, your stomach churns away to start the digestion process. This could be why you can hear rumbling stomachs in quiet rooms.

Act or relax?

Automatic, unconscious functions of the body are managed by the "primitive" parts of the central nervous system – the spinal cord and brainstem. However, they use two different networks of nerves to control our body parts depending on whether we need to get moving or put our feet up.

Calming the nerves

Our twin automatic nervous systems are called the sympathetic and the parasympathetic. Together they form what is called the autonomic nervous system. The parasympathetic nerves tend to slow things down and start digesion. You don't tend to notice their effects.

Bladder contracts
You regain complete control over the bladder muscles. They keep your bladder shut when you are fully relaxed.

Intestine speeds up
Nutrients are absorbed from the small intestines, and bowel movements push undigested waste onwards. This process works best when you are still and relaxed.

Digestion slows
Your stomach is instructed to bring digestion to a halt. In times of true terror, you may vomit to stop digestion. A full stomach can slow you down if running.

Intestine slows
Blood is diverted from the intestine, since it is an unimportant organ in times of stress, and movements in your gut slow down or stop altogether.

Bladder relaxes
The muscles that usually keep the bladder shut tend to relax if you are anxious, unfortunately resulting in frequent trips to the toilet.

Braced for action

The job of igniting and stimulating your body ready for action lies with the sympathetic nervous system, which uses different nerves. Once it has served its purpose, the parasympathetic system kicks in, counteracting the sympathetic effects to wind your body back down into a relaxed state.

BUTTERFLIES IN THE STOMACH

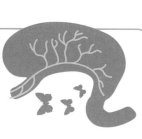

The sensation of butterflies before a stage performance or big interview is due to the reduction of blood flow to the stomach when readying the body for danger. The stomach has a dense network of nerves, and some of these nerves signal nervous, fluttery feelings, or even nausea, as blood flow drops.

Knocks, sprains, and tears

Soft tissues of the body, such as nerves, muscles, tendons, and ligaments, are susceptible to injury, leading to bruising, swelling, inflammation, and pain. Some injuries result from sports, while others can occur from over-use or accidents. Injuries are more common with age and poor fitness.

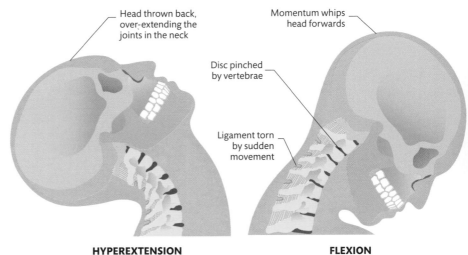

WHY DOES HITTING YOUR "FUNNY BONE" FEEL FUNNY?

Knocking your elbow compresses the ulnar nerve, which runs down the outside of your elbow, against bone, causing an electric shock sensation.

Nerve problems

Nerves stretch for long distances and often travel through narrow spaces between bones. These tunnels guide and protect the nerve, but can also trap it to cause pain, numbness, or tingly feelings. Pinching can occur when repetitive movements cause tissues to swell, from maintaining an awkward position for a long time (such as keeping an elbow bent during sleep), or when surrounding tissues move out of alignment, which occurs with a slipped disc.

Carpal ligament

Muscles further up the arm shield nerves from potential knocks or pressure

Median nerve

Ulnar nerve

Exposed ulnar nerve which is where you may hit your "funny bone"

Elbow

Carpal tunnel syndrome
The median nerve passes between the wrist bones and a strong ligament connecting the base of the thumb and little finger. Pinching of the nerve causes painful tingling in the hand, wrist, and forearm.

Whiplash

This injury to the neck occurs when the head is suddenly whipped backwards and then forwards or vice versa. This commonly happens to those travelling in a car that is hit from behind by another vehicle.

Head thrown back, over-extending the joints in the neck

Momentum whips head forwards

Disc pinched by vertebrae

Ligament torn by sudden movement

Squashed discs and torn ligaments
The sudden whiplash movement jars the neck. This motion can injure bones in the spine, compress discs between the vertebrae, tear ligaments and muscles, and stretch nerves in the neck.

HYPEREXTENSION

FLEXION

Back pain

Back pain most commonly occurs in the lower spine, which is vulnerable as it supports most of the body's weight. Many cases result from heavy lifting without protecting the back by keeping it straight. Excessive strain can lead to tearing and spasms of the muscles, the stretching of ligaments, and even a dislocation of one of the tiny gliding joints (see p.40) between the vertebrae. Pressure may cause the soft, jelly-like centre of an intervertebral disc to rupture through its fibrous coat and press on a nerve. Treatment involves painkillers, manipulation, and remaining as mobile as possible.

Muscle tears in your back are difficult to heal as blood flow is limited

Muscle strain
When you are unfit, muscles have poor tone. They are easily strained from lifting, carrying, bending awkwardly, or even prolonged sitting in one position.

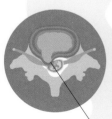

Slipped disc
A damaged spinal disc presses on a nerve root causing pins and needles, spasm, and back pain. Sciatic nerve irritation causes shooting pain down one leg.

Slipped spinal disc

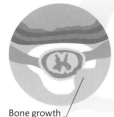

Bone spurs
As ageing vertebrae start to wear out, mild inflammation and the bone's attempt to heal can produce spur-like growths that press against nerve roots causing pain.

Bone growth

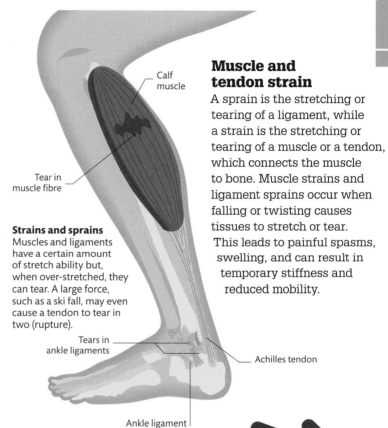

Calf muscle

Tear in muscle fibre

Strains and sprains
Muscles and ligaments have a certain amount of stretch ability but, when over-stretched, they can tear. A large force, such as a ski fall, may even cause a tendon to tear in two (rupture).

Tears in ankle ligaments

Ankle ligament

Achilles tendon

Muscle and tendon strain

A sprain is the stretching or tearing of a ligament, while a strain is the stretching or tearing of a muscle or a tendon, which connects the muscle to bone. Muscle strains and ligament sprains occur when falling or twisting causes tissues to stretch or tear. This leads to painful spasms, swelling, and can result in temporary stiffness and reduced mobility.

THE **ANKLE** IS THE **MOST COMMON** AREA OF THE BODY **TO GET A SPRAIN.**

"PRICE" TECHNIQUE

The PRICE technique is an effective way to treat a strain or sprain: Protection – use a support, crutch, or sling to relieve pressure. Rest – keep the injured area free from movement. Ice – apply an ice-pack to minimize swelling and bleeding. Compression – an elasticated bandage reduces swelling. Elevation – keep the area raised to reduce swelling.

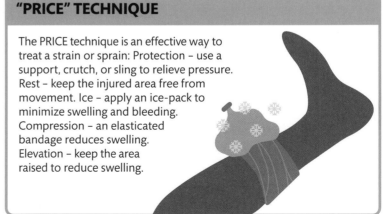

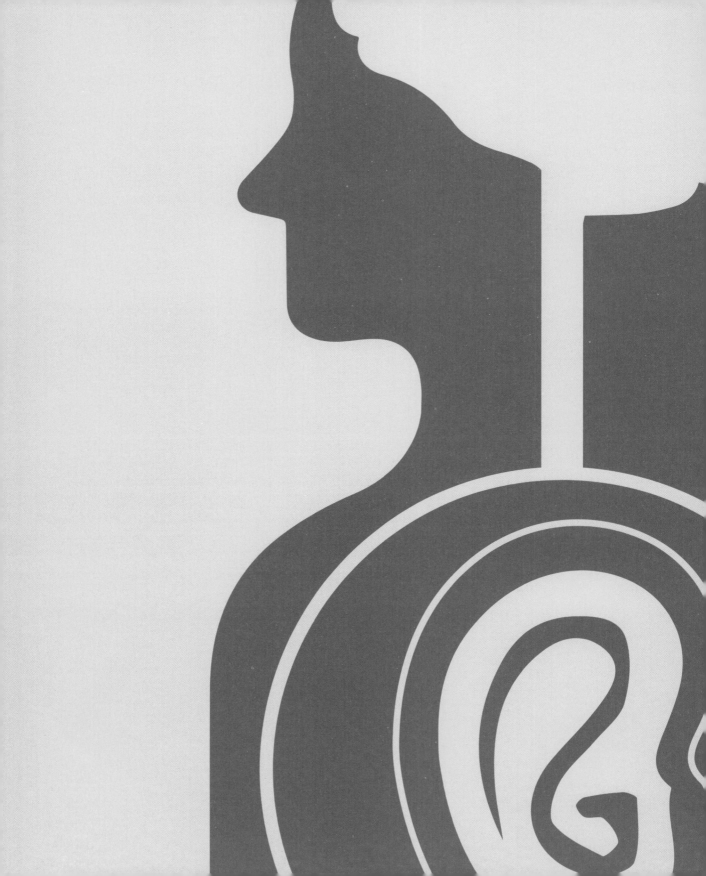

SENSITIVE TYPES

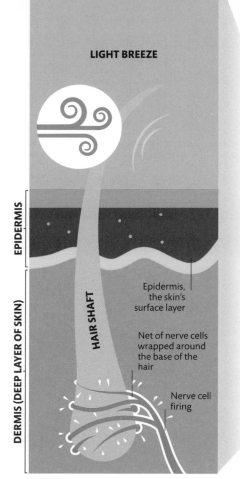

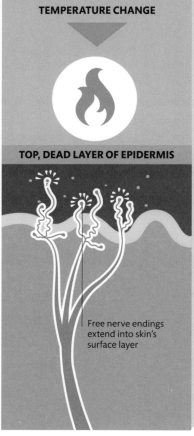

LIGHT BREEZE

TEMPERATURE CHANGE

BRUSH OF A FEATHER

EPIDERMIS

DERMIS (DEEP LAYER OF SKIN)

TOP, DEAD LAYER OF EPIDERMIS

HAIR SHAFT

Epidermis, the skin's surface layer

Net of nerve cells wrapped around the base of the hair

Nerve cell firing

Free nerve endings extend into skin's surface layer

Very light touch receptors rest against the base of the epidermis

Hair movement
We can sense things that haven't touched our skin. Air currents, or the brushing of hair against objects, distorts and triggers nerves wrapped around a hair's base.

Temperature and pain
Nerves without any special structure around them are sensitive to cold, heat, or pain. They are the shallowest receptors, extending right into the skin's surface layer.

Very light touch
Slightly lower than the free nerve endings are Merkel's cells, which are sensitive to the faintest touch. They are particularly dense in the fingertips.

Feeling the pressure

What we think of as our sense of touch is actually composed of signals from several different receptors in our skin. Some receptors are concentrated in certain areas, such as the sensitive fingertips.

How the skin feels

Our skin is full of microscopic sensors, or receptors, that are buried at different depths and are poised to respond to touches of different kinds – from faint, brief contacts to sustained pressure. In effect, each represents a subtly distinct sense. Receptors work by responding (triggering a nerve impulse) when they are disturbed or distorted.

HOW DO WE FEEL DEEP INSIDE THE BODY?

Nearly all of our touch sense is in the skin and joints. But we also feel discomfort in our guts. This comes from stretch receptors and chemical sensors in and around our intestines.

GENTLE TOUCH

Light touch receptors sit at the top of the dermis

Light touch
Light-touch receptors are good for reading Braille, because they are arranged densely and their firing dies away quickly. This gives precise, rapidly updating information.

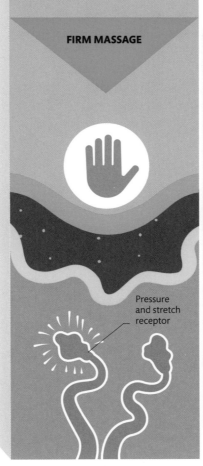

FIRM MASSAGE

Pressure and stretch receptor

Pressure and stretch
If the skin is stretched or distorted by pressure, deep receptors fire. They stop firing after a few seconds, so they report rapid changes, not continuous pressure.

VIBRATION

Deep pressure and vibration receptor

Vibration and pressure
The deepest type of touch receptor occurs in joints as well as skin. These sensors don't give up firing, so they respond to sustained pressure, as well as vibration.

FROM PALM TO FINGERTIP

Our palms and fingers are very sensitive, but our fingertips have more nerve endings than anywhere else on our skin. Light-touch sensors are packed by the thousand into the pads of our fingers. The pattern in which they fire tell us about the texture of surfaces that we touch.

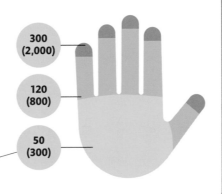

300 (2,000)

120 (800)

50 (300)

Number of nerve endings per sq cm (per sq in)

EACH OF YOUR **FINGERTIPS** CAN DETECT DIFFERENCES IN TEXTURE **10,000 TIMES SMALLER** THAN THE **WIDTH OF A HAIR**

How do you feel?

From our skin, tongue, throat, joints, and other body parts, microscopic sensors send touch information along sensory nerves to the brain. The destination of these nerve impulses is a part of the brain's outer layer called the sensory cortex, where the touch information is organized and analyzed.

Homunculus
A sensory homunculus is a body pictured in proportion to the area of sensory cortex devoted to it. The colours of this one match those on the large illustration of the brain.

How the brain feels

We can tell where something touches us because the brain contains a map of the body. The map is on a strip of the brain's outer layer called the sensory cortex, but it is distorted. Because some body parts are so much more sensitive, with closely packed nerve endings, those parts occupy a hugely exaggerated area of the map. The cortex needs such a great area to record precisely the detailed touch data. It combines the information to calculate whether an object is hard or soft, rough or smooth, warm or cold, stiff or flexible, wet or dry, and much more.

Touch-sensitive brain
Viewed from the side, the part of the brain's surface that receives touch information is a narrow strip. It continues down the inside into the deep canyon between the brain's two halves.

SENSORY CORTEX

CORTEX

This pink band is the sensory cortex – the part of the cortex that receives touch information

The cortex, in yellow, is the outer layer of the cerebrum – the giant, folded structure that forms most of the human brain

Sensitive bits
The cortex reserves a disproportionate amount of space for the body parts that deliver the most detailed touch information – the lips, palms, tongue, thumb, and fingertips.

5 MILLION
THE TOTAL AMOUNT OF **SENSORY NERVE** ENDINGS IN **THE SKIN**

LEFT HEMISPHERE receives touch information from the right side of the body

HOW DO WE SENSE TEMPERATURE?

Specific skin nerve endings are sensitive to hot or cold. In the range 5–45°C (41–113°F), both types fire all the time, but at different rates, giving the brain an idea of how hot or cold it is. Outside this range, different nerve endings take over. These register not heat, but pain.

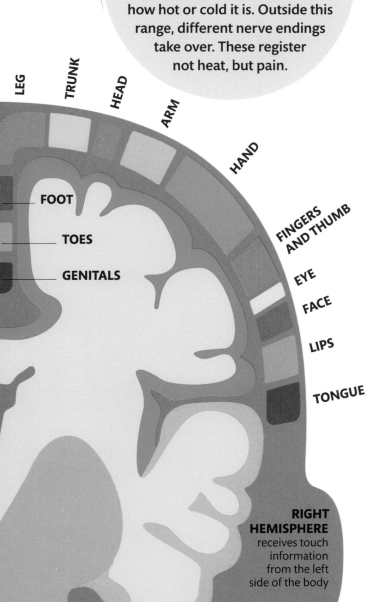

LEG
TRUNK
HEAD
ARM
HAND

FOOT
TOES
GENITALS

FINGERS AND THUMB
EYE
FACE
LIPS
TONGUE

RIGHT HEMISPHERE receives touch information from the left side of the body

Why can't we tickle ourselves?

When we try to tickle ourselves, our brain takes a copy of the intended movement pattern of our fingers and sends it to the body part about to be tickled, warning it and dampening its tickle response. This works because unlike tickles from other people, our brain can predict the precise movement of our own hands and filter it out. This is an example of the brain's vital ability to filter unwanted sensory data.

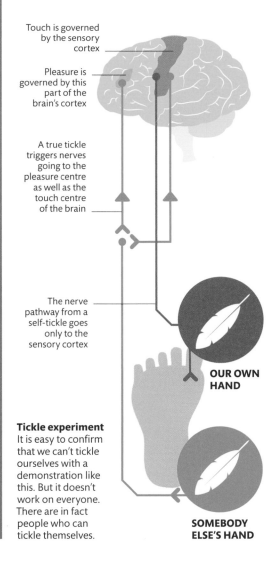

Touch is governed by the sensory cortex

Pleasure is governed by this part of the brain's cortex

A true tickle triggers nerves going to the pleasure centre as well as the touch centre of the brain

The nerve pathway from a self-tickle goes only to the sensory cortex

OUR OWN HAND

Tickle experiment
It is easy to confirm that we can't tickle ourselves with a demonstration like this. But it doesn't work on everyone. There are in fact people who can tickle themselves.

SOMEBODY ELSE'S HAND

Pain's pathway

Pain, while unpleasant, is actually incredibly helpful. It tells you when your body is damaged, and the level of pain you feel helps you act accordingly.

Feeling the pain

Pain signals travel from nerve cell receptors at the site of injury along nerves to the spinal cord, and then to the brain, which tells you that you are in pain. Man-made or natural painkilling chemicals work by stopping this flow of information.

Slow C-fibre

Fast A-fibre

Myelin sheath

NERVE BUNDLE

Blocked at the nerve
Local anaesthetic blocks conduction of electrical impulses along the A and C nerve fibres, so these impulses never reach the spinal cord.

3 Fast or slow?
A-fibre axons are wrapped in myelin sheaths, allowing electrical signals to travel faster than C-fibres. Dense A-fibre receptors in the skin result in sharp, localized pain. Slower C-fibres produce dull, burning aches.

DULL, GENERAL ACHE

SHARP, LOCALIZED PAIN

PAIN SIGNALS TRAVEL UP TO 15 TIMES FASTER ALONG A-FIBRES THAN C-FIBRES

2 Stimulated nerve cell
Exposed nerve endings in your skin start to fire in response to prostaglandins. Electrical signals signalling pain are carried by nerve cell axons into nerve bundles.

Axon

Nerve cell

Blocked at injury
Aspirin blocks generation of prostaglandins at the site of injury to stop nerve sensitization.

1 Prostaglandins
When you hurt yourself, cells in your skin are damaged. Damaged cells release chemicals called prostaglandins which sensitize surrounding nerve cells.

Prostaglandin molecule released by cell

Damaged cell

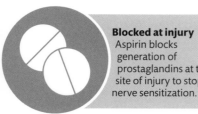

Physical damage directly stimulates pain receptors giving us our first sensation of pain when injured

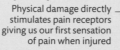

SKIN

BRUISE

CUT

REFERRED PAIN

Nerve pathways from our internal organs run alongside nerve pathways from the skin and muscles before reaching our brain. This means the brain may misinterpret pain from the organ as occurring in the nearby muscles or skin, which is more common and likely.

Heart pain signal

Feeling of pain felt on arm and right side of chest

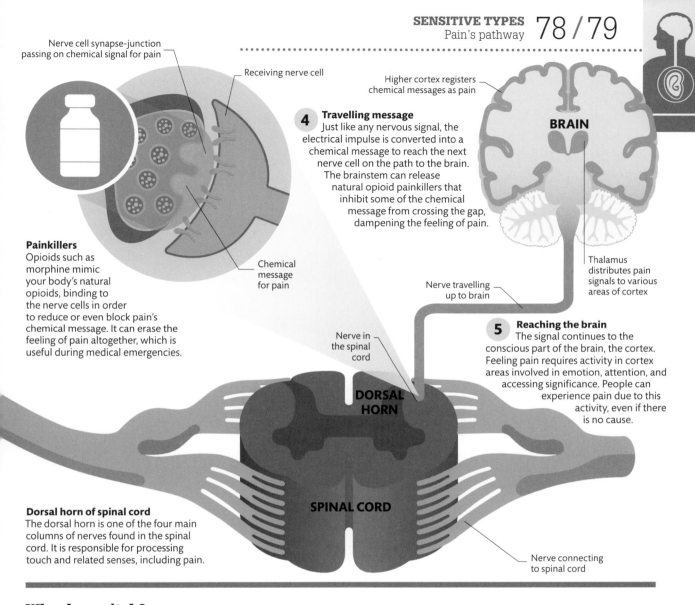

Nerve cell synapse-junction
passing on chemical signal for pain

Receiving nerve cell

Higher cortex registers
chemical messages as pain

BRAIN

4 Travelling message
Just like any nervous signal, the
electrical impulse is converted into a
chemical message to reach the next
nerve cell on the path to the brain.
The brainstem can release
natural opioid painkillers that
inhibit some of the chemical
message from crossing the gap,
dampening the feeling of pain.

Painkillers
Opioids such as
morphine mimic
your body's natural
opioids, binding to
the nerve cells in order
to reduce or even block pain's
chemical message. It can erase the
feeling of pain altogether, which is
useful during medical emergencies.

Chemical
message
for pain

Thalamus
distributes pain
signals to various
areas of cortex

Nerve travelling
up to brain

Nerve in
the spinal
cord

5 Reaching the brain
The signal continues to the
conscious part of the brain, the cortex.
Feeling pain requires activity in cortex
areas involved in emotion, attention, and
accessing significance. People can
experience pain due to this
activity, even if there
is no cause.

**DORSAL
HORN**

SPINAL CORD

Dorsal horn of spinal cord
The dorsal horn is one of the four main
columns of nerves found in the spinal
cord. It is responsible for processing
touch and related senses, including pain.

Nerve connecting
to spinal cord

Why do we itch?

Itches arise when our skin is irritated by
something on its surface, or by chemicals
released by the body when parts of the skin have
become inflamed due to disease. It is likely to
have evolved to protect us against biting insects.
Itch receptors are separate from touch or pain
receptors. When they are stimulated, a signal
travels through the spinal cord to the brain
where the scratch response is initiated.
Scratching an itch stimulates both touch and
pain receptors, blocking signals from the itch
receptor and distracting you from the urge to itch.

ITCH

Scratching
an itch

PAIN

RELIEF

Itching cycle
Scratching can irritate
the skin further, which
makes the itch signal
ever more persistant.
Scratching also causes
the brain to release
serotonin to dampen the
pain caused, providing
temporary relief. Though
once this wears off, the
urge to itch returns
stronger than before.

How the eye works

Our visual capabilities are amazing. We can see detail and colour, see near and far objects clearly, and judge speed and distance. The first stage in the visual process is image capture – a sharp image forms on the eye's light receptors. The image then needs to be converted into nerve signals (see pp.82–83) so that it can be processed by the brain (see pp.84–85).

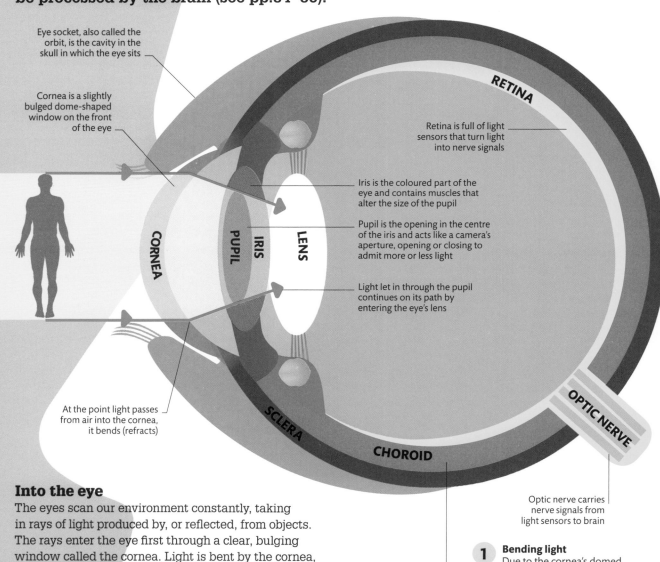

Eye socket, also called the orbit, is the cavity in the skull in which the eye sits

Cornea is a slightly bulged dome-shaped window on the front of the eye

RETINA

Retina is full of light sensors that turn light into nerve signals

Iris is the coloured part of the eye and contains muscles that alter the size of the pupil

Pupil is the opening in the centre of the iris and acts like a camera's aperture, opening or closing to admit more or less light

Light let in through the pupil continues on its path by entering the eye's lens

CORNEA

PUPIL

IRIS

LENS

At the point light passes from air into the cornea, it bends (refracts)

SCLERA

CHOROID

OPTIC NERVE

Optic nerve carries nerve signals from light sensors to brain

Into the eye

The eyes scan our environment constantly, taking in rays of light produced by, or reflected, from objects. The rays enter the eye first through a clear, bulging window called the cornea. Light is bent by the cornea, passing through the pupil – which controls light intensity – and is then fine-focused by the adjustable lens onto the retina, whose millions of photoreceptor cells form an image to be sent to the brain.

Choroid contains blood vessels that supply the retina and sclera with blood

1 Bending light
Due to the cornea's domed shape, light refracting through it bends inwards through the pupil towards a focal point within the eye. The pupil, which is a hole in the iris, lets a controlled amount of light through.

Ciliary muscles contract to make lens fatter for close focus, or relax to make it thinner for distant objects

Iris

Ligaments attach cilary muscle to lens

Lens is elastic and gets rounder when ligaments are slack

Optic nerve

2 Autofocusing
As we look at nearby and distant objects, we adjust the focus of our eyes without thinking. For close work, the muscles that pull on the lens contract, the ligaments go slack, and the lens bulges to increase its focusing power.

Light sensors in retina send nerve signals in response to image

Image on retina is upside down

Optic nerve carries nerve signals to brain

3 Image on the retina
When light hits the retina, more than 100 million light receptors are stimulated, like the pixels on a digital camera's sensor. The pattern of light intensity and colour in the image is preserved as an electrical signal in the optic nerve, which sends it to the brain.

Bright light

The iris is the coloured part of the eye with a central opening called the pupil. It contains muscles that contract or relax to alter the size of the pupil and so let more or less light into the eye.

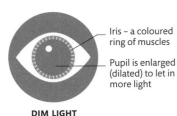

Iris – a coloured ring of muscles

Pupil is enlarged (dilated) to let in more light

DIM LIGHT

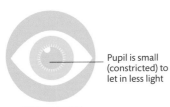

Pupil is small (constricted) to let in less light

BRIGHT LIGHT

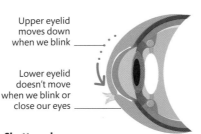

Upper eyelid moves down when we blink

Lower eyelid doesn't move when we blink or close our eyes

Shutters down
Our eyes are extremely delicate. The eyelids close by reflex action if we are in danger of getting something in our eyes.

First line of defence

The eyelashes and eyelids help to protect our eyes. The eyelashes prevent dust and other small particles from getting into the eyes. The eyelids help to protect against larger objects and irritant substances in the air. The eyelids also spread tears across the surface of the eye.

Lubrication
Produced by tear glands under the upper eyelid, tears moisten and lubricate the eye and wash away small particles from the eye's surface. Tears are produced continually, although we only notice when we cry or our eyes water.

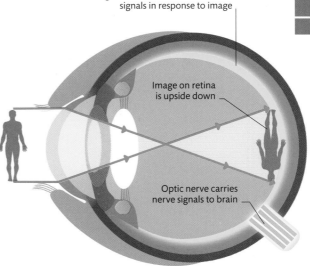

Tear gland produces tears, which trickle into the eye through tear ducts

Tear drops form when the tear glands produce too much tear fluid for it to drain away through the nose

Channel drains tears into the nose

Forming an image

The part of our eye that creates images, the retina, is only the size of a thumbnail, but can produce an incredibly sharp and detailed image. We rely on cells inside the retina to convert light rays into images.

How we see

Images are formed at the back of the eye in a layer called the retina. Cells inside the retina are sensitive to light. When light rays strike them, they trigger nerve signals, which then travel to the brain to be processed as an image. The retina contains two types of light sensor cells; cone cells, or cones, detect colour (wavelength) of light rays, whereas rod cells, or rods, do not.

WHAT ARE LIGHT SPOTS?

The gel-like fluid that fills the inner part of your eye can break loose, blocking incoming light rays and casting shadows on your retina. These shadows appear as flashing dots or shapes in your vision.

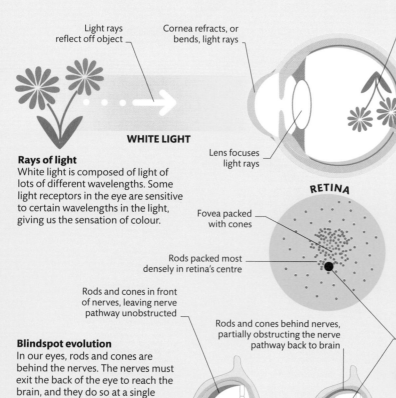

Light rays reflect off object

Cornea refracts, or bends, light rays

Inverted object

Lens focuses light rays

WHITE LIGHT

RETINA

Fovea packed with cones

Rods packed most densely in retina's centre

Rays of light
White light is composed of light of lots of different wavelengths. Some light receptors in the eye are sensitive to certain wavelengths in the light, giving us the sensation of colour.

Rods and cones
Rods are packed most densely around the centre of the retina, although none are found in the central region, known as the fovea. The fovea is packed with cones, and there are no blood vessels in this small area, so it produces a sharp, detailed picture. The very centre of the fovea contains only red and green cones.

Rods and cones in front of nerves, leaving nerve pathway unobstructed

Rods and cones behind nerves, partially obstructing the nerve pathway back to brain

Blind spot where optic nerve reaches back of eye

Blindspot evolution
In our eyes, rods and cones are behind the nerves. The nerves must exit the back of the eye to reach the brain, and they do so at a single point, creating a blind spot with no rods or cones. Our brain compensates by guessing what should be in the blank region and filling it in for us. On the other hand, the eyes of squid have nerves that sit behind their rods and cones, resulting in no blind spot.

SQUID EYE

HUMAN EYE

20-100
MILLISECONDS - THE TIME TAKEN FOR YOUR EYES TO MAKE A MOVEMENT WHEN **READING QUICKLY.**

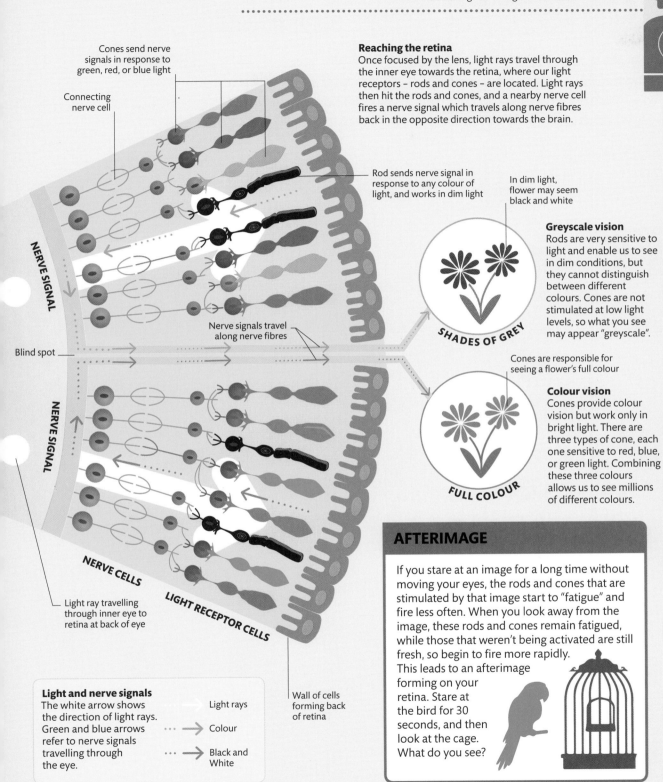

Cones send nerve signals in response to green, red, or blue light

Connecting nerve cell

NERVE SIGNAL

Blind spot

NERVE SIGNAL

NERVE CELLS

LIGHT RECEPTOR CELLS

Light ray travelling through inner eye to retina at back of eye

Rod sends nerve signal in response to any colour of light, and works in dim light

Nerve signals travel along nerve fibres

Reaching the retina
Once focused by the lens, light rays travel through the inner eye towards the retina, where our light receptors – rods and cones – are located. Light rays then hit the rods and cones, and a nearby nerve cell fires a nerve signal which travels along nerve fibres back in the opposite direction towards the brain.

In dim light, flower may seem black and white

SHADES OF GREY

Greyscale vision
Rods are very sensitive to light and enable us to see in dim conditions, but they cannot distinguish between different colours. Cones are not stimulated at low light levels, so what you see may appear "greyscale".

Cones are responsible for seeing a flower's full colour

FULL COLOUR

Colour vision
Cones provide colour vision but work only in bright light. There are three types of cone, each one sensitive to red, blue, or green light. Combining these three colours allows us to see millions of different colours.

Light and nerve signals
The white arrow shows the direction of light rays. Green and blue arrows refer to nerve signals travelling through the eye.

→ Light rays
···→ Colour
···→ Black and White

Wall of cells forming back of retina

AFTERIMAGE

If you stare at an image for a long time without moving your eyes, the rods and cones that are stimulated by that image start to "fatigue" and fire less often. When you look away from the image, these rods and cones remain fatigued, while those that weren't being activated are still fresh, so begin to fire more rapidly. This leads to an afterimage forming on your retina. Stare at the bird for 30 seconds, and then look at the cage. What do you see?

Vision in the brain

Our eyes provide basic visual data about the world, but it is our brain that extracts useful information from it. This is done by selectively modifying it, producing our visual perception of the world – deducing movement and depth and taking into account lighting conditions.

Binocular vision

We are able to see in 3-D because of the placement of our eyes. They both point in the same direction, but are spaced apart slightly, so that they see slightly different images when looking at an object. How different these images are depends on the distance of the object relative to where you are fixating, so we use the disparity between the images to judge how far away an object is.

Visual pathways
Information from the eyes is carried to the back of the brain, where it is processed and turned into conscious vision. Along the way, signals converge at the optic chiasm, where half of the signals cross over to the opposite hemisphere of the brain.

VISUAL FIELD OF THE LEFT EYE

BINOCULAR VISUAL FIELD

This is the image formed by the brain after it combines the images from the left and right eyes' visual fields

VISUAL FIELD OF THE RIGHT EYE

Seeing in 3-D

The way our brains have evolved to perceive depth can be used to produce 3-D movies. Filmmakers film one image out of polarized light waves that are oscillating up and down, and a different image, filmed from a different angle, from light oscillating from side to side. By providing each eye with these slightly different images, they trick the brain into thinking it is seeing in 3-D.

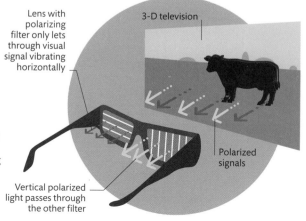

Lens with polarizing filter only lets through visual signal vibrating horizontally

3-D television

Polarized signals

Vertical polarized light passes through the other filter

24

THE NUMBER OF **FRAMES PER SECOND** AT WHICH **FILM IS RECORDED**

Perspective

Experience tells us that two straight lines, such as railway tracks, appear to converge in the distance. We use this to estimate depth from an image – by combining this with other cues, such as changes in texture and comparisons to objects of known size, we can estimate distances. The image to the right creates an illusion because we interpret converging lines as distance and compare the cars' sizes to lane width.

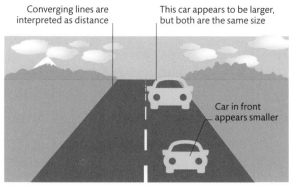

Converging lines are interpreted as distance

This car appears to be larger, but both are the same size

Car in front appears smaller

PERSPECTIVE ILLUSION

LEFT HEMISPHERE

LEFT OPTIC TRACT

THALAMUS

OPTIC CHIASM

LEFT VISUAL CORTEX

RIGHT VISUAL CORTEX

THALAMUS

RIGHT OPTIC TRACT

RIGHT HEMISPHERE

Left visual cortex receives signals from left side of each retina

Right visual cortex receives signals from right side of each retina

The right optic radiation is a band of nerve fibres that carries the visual signal from the thalamus to the right visual cortex

COLOUR CONSTANCY

We are used to seeing objects in a variety of lighting conditions and our brain takes this into account to cancel out the effects of shadows and lighting. This means we always see a banana as yellow, no matter how it is illuminated. But sometimes our brains see only what they expect.

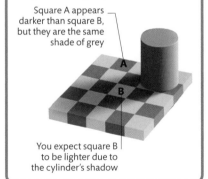

Square A appears darker than square B, but they are the same shade of grey

You expect square B to be lighter due to the cylinder's shadow

Moving pictures

Surprisingly, our eyes don't provide a smooth stream of moving visual information. They deliver a series of snapshots to the brain, just like film or video. The brain creates the perception of movement from the images, which is why we find it easy to blend the frames of film and TV into the impression of smooth motion. The process can go wrong, however, because a sequence of still frames can be misleading.

FRAME 1   FRAME 2   Real motion between frames   Perceived motion between frames

FRAME 3   FRAME 4

Apparent motion
When the wheels of cars on TV seem to go backwards, it is because they make a little less than one rotation between frames. Our brain wrongly reconstructs a slow backward motion.

Eye problems

Your eyes are complex, delicate organs and therefore vulnerable to disorders caused by damage or natural degeneration as you get older. Eye problems affect most people at some point in their lives, but luckily, many eye conditions are easily treatable.

Why do you need glasses?

You see sharp, clear images when light from an object is bent by your lens and cornea and focused on the retina (see pp.80–81). If this system is slightly off, images appear blurred. Glasses can correct for too much or too little bending of the light, bringing the image back into focus. The prevalence of short-sightedness appears to be increasing – possibly because modern day life, especially in urban environments, requires us to focus more on objects nearby than those far away.

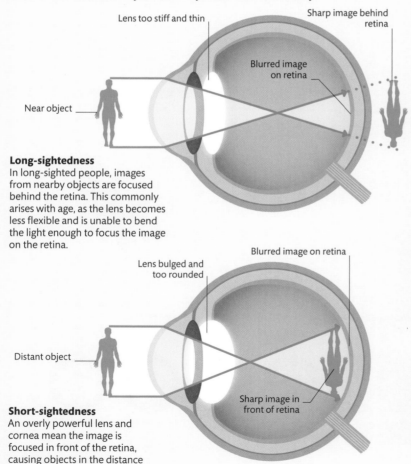

Lens too stiff and thin

Sharp image behind retina

Blurred image on retina

Near object

Long-sightedness
In long-sighted people, images from nearby objects are focused behind the retina. This commonly arises with age, as the lens becomes less flexible and is unable to bend the light enough to focus the image on the retina.

Lens bulged and too rounded

Blurred image on retina

Distant object

Sharp image in front of retina

Short-sightedness
An overly powerful lens and cornea mean the image is focused in front of the retina, causing objects in the distance to appear blurry.

90%

THE PROPORTION OF 16–18 YEAR OLD CHILDREN WITH SHORT-SIGHTEDNESS IN SOME CITIES.

Astigmatism

The most common type of astigmatism is caused by a cornea or lens shaped more like a rugby ball than a football. This means that while the image may be focused on the retina horizontally, the vertical aspect could be focused in front of or behind the retina (or vice versa). It can be corrected using glasses or contact lenses, or through laser eye surgery.

What you see
People with astigmatism may see vertical or horizontal lines blurred, but the other in focus. Sometimes, both axes are distorted – one can be long-sighted and the other short-sighted.

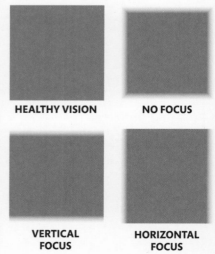

HEALTHY VISION	**NO FOCUS**
VERTICAL FOCUS	**HORIZONTAL FOCUS**

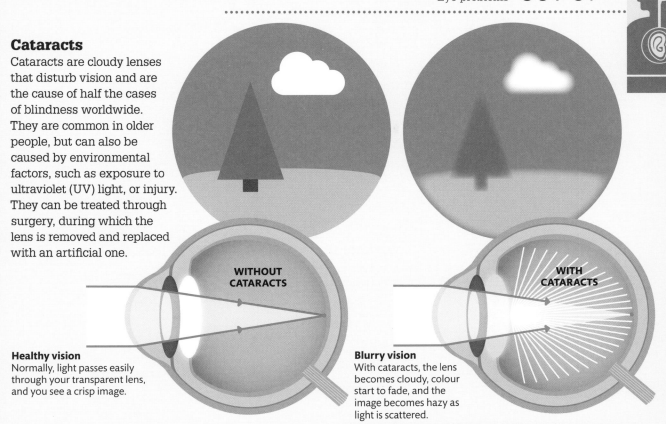

Cataracts

Cataracts are cloudy lenses that disturb vision and are the cause of half the cases of blindness worldwide. They are common in older people, but can also be caused by environmental factors, such as exposure to ultraviolet (UV) light, or injury. They can be treated through surgery, during which the lens is removed and replaced with an artificial one.

WITHOUT CATARACTS

WITH CATARACTS

Healthy vision
Normally, light passes easily through your transparent lens, and you see a crisp image.

Blurry vision
With cataracts, the lens becomes cloudy, colour start to fade, and the image becomes hazy as light is scattered.

Glaucoma

Normally, excess fluid in your eyes drains harmlessly into the blood. Glaucoma occurs when blocked drainage channels cause fluid to build-up in the eye. Causes of glaucoma aren't well understood, although genetics plays a part.

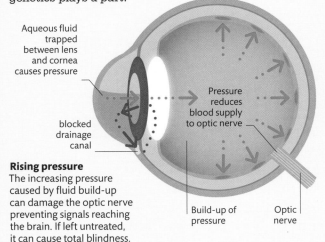

Aqueous fluid trapped between lens and cornea causes pressure

blocked drainage canal

Pressure reduces blood supply to optic nerve

Rising pressure
The increasing pressure caused by fluid build-up can damage the optic nerve preventing signals reaching the brain. If left untreated, it can cause total blindness.

Build-up of pressure

Optic nerve

TESTING YOUR VISION

Vision tests allow optometrists to examine your ability to see at long and short distances and to check that your eyes are working together and the muscles are healthy. They also inspect the eye inside and out, which can pick up illnesses such as diabetes and vision problems such as glaucoma or cataracts. Another type of vision problem that may be detected is colour blindness. Colour vision deficiencies are caused by missing or defective cone types, so sufferers rely on fewer than the three cone types most people have. This means they confuse certain colours – most commonly reds and greens.

Some people see the number 74, some 21, and some neither

How the ear works

Our ears have the tricky job of converting sound waves in the air into nerve signals for our brains to interpret. The series of steps used ensures as much of the information as possible is preserved. Ears can also amplify faint signals, and determine where sounds are coming from.

Getting sound into the body

When sound waves travel from air to liquid, as they must to enter the body, they are partially reflected, so they have less energy and sound quieter. Our ear prevents the sound bouncing off by easing the wave energy in, step by step. When the ear drum vibrates, it pushes on the first of three tiny bones called ossicles, which move in turn, pushing on the oval window and setting up waves in the cochlea's liquid. As the sound passes through the ossicles, they amplify it by 20–30 times.

Easing sound in

Sound waves travel down the ear canal and cause the ear drum to vibrate. The vibration is passed through the three ossicles. Due to the way they pivot, they use leverage to amplify the vibration in steps. The last ossicle pushes at the oval window – the entrance to the inner ear, where the vibrations pass into the fluid of the cochlea.

The three semicircular canals in the inner ear are balance organs and not part of hearing

SEMICIRCULAR CANAL

Malleus (hammer) bone is the first of the ear ossicles

INNER EAR

OSSICLES

Vibration passes from ear drum to malleus bone

Ear drum vibrates

MIDDLE EAR

Oval window – a membrane, like the ear drum

Incus (anvil) bone passes vibration to the final ossicle, the stapes

Stapes (stirrup) bone pushes fluid in the cochlea through a membrane-covered window

OUTER EAR

EAR CANAL

PINNA (EXTERNAL EAR)

Sound vibrations enter ear canal

Shape of external ear, or pinna, funnels sound waves into ear canal and gives clues about whether they came from in front or behind

WHY DON'T OUR OWN VOICES DEAFEN US?

Our ears are less sensitive when we speak, because tiny muscles hold the ossicles steady, dampening their vibration. Less energy is passed into the cochlea and it causes no damage.

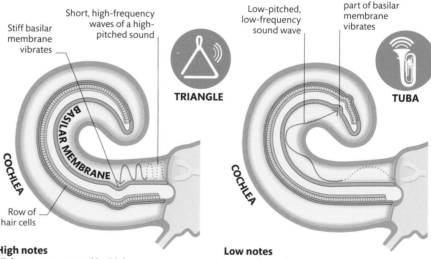

VESTIBULAR NERVE

AUDITORY NERVE

Auditory nerve sends electrical signals to the brain

COCHLEA

Sound passes through the fluid of the cochlea

Eustachian tube connects ear to nose and mouth

EUSTACHIAN TUBE

THE WORD COCHLEA COMES FROM THE GREEK FOR SNAIL, BECAUSE OF ITS COILED SHAPE

Sounds of different pitches

Inside the cochlea is the basilar membrane, which is connected to sensitive hair cells. Each section of the membrane vibrates most at a particular frequency, because its stiffness changes along its length. Different sounds therefore cause deflection of different hair cells. The brain deduces the pitch of the sound using the position of the disturbed cells.

Stiff basilar membrane vibrates

Short, high-frequency waves of a high-pitched sound

TRIANGLE

Low-pitched, low-frequency sound wave

More flexible part of basilar membrane vibrates

TUBA

BASILAR MEMBRANE

COCHLEA

Row of hair cells

COCHLEA

High notes
High notes are caused by high-frequency waves. These activate the basilar membrane near its base, where it is narrower and stiffer, and vibrates more rapidly.

Low notes
Longer, lower-frequency waves travel further through the cochlea before causing the basilar membrane to vibrate nearer its tip, where it is floppier and wider.

Sound into electricity

The information in the sound – including its pitch, tone, rhythm, and intensity – is converted into electrical signals to be sent to the brain for analysis. Exactly how the information is encoded is still unknown, but it is achieved by the hair cells and auditory nerves.

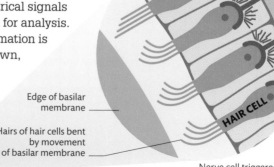

Edge of basilar membrane

Hairs of hair cells bent by movement of basilar membrane

HAIR CELL

Nerve cell triggered to send signal to the brain

LOCATION IN THE COCHLEA

Triggering the nerves
When the sensitive hairs on the hair cells are moved by vibration of the basilar membrane, they release neurotransmitters that trigger nerve cells at their bases.

How the brain hears

Once signals from the ear reach the brain, complex processing is needed to extract information. Our brains determine what the sound is, where it is coming from, and how we feel about it. The brain is able to focus in on one sound over another, and even tune out unnecessary noises completely.

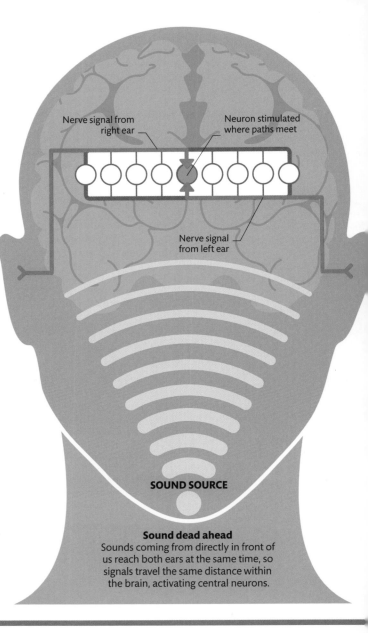

Nerve signal from right ear

Neuron stimulated where paths meet

Nerve signal from left ear

SOUND SOURCE

Sound dead ahead
Sounds coming from directly in front of us reach both ears at the same time, so signals travel the same distance within the brain, activating central neurons.

Localizing sound

We use three main cues to find where a sound is coming from – its frequency pattern, loudness, and the difference in arrival time at each ear. We use frequency pattern to tell if the sound is in front or behind us, because our ear's shape means that a sound coming from in front has a different pattern of frequencies than the same sound coming from behind. Our ears don't help much in pinning down the height of sound sources, though. Left and right localization is easier – a sound from the left is louder in the left ear than in the right, particularly at high frequencies. It also reaches our left ear a few milliseconds before our right. The diagrams on the right show how the brain uses this information.

Tuned in

Our brains can "tune in" to a single conversation over the babble of noise at a party by grouping sounds into separate streams, based on frequency, timbre, or source. It might seem as though you don't hear any of the other conversations – but you will notice if someone mentions your name. That's because your ears still send signals from the other conversations to the brain, which will override the filtering if something important comes up elsewhere.

WE CAN PICK OUT A CONVERSATION IN NOISY ENVIRONMENTS

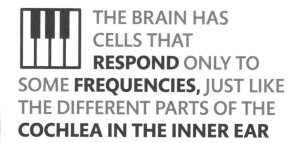

THE BRAIN HAS CELLS THAT **RESPOND** ONLY TO SOME **FREQUENCIES**, JUST LIKE THE DIFFERENT PARTS OF THE **COCHLEA IN THE INNER EAR**

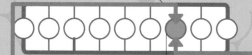

Signal travels further from this side before meeting the pathway from other ear

The neuron that fires tells us how far to the left or right the sound is coming from

Sound waves reach the nearer ear first

Off-centre sound source
Different neurons are activated depending on the delay between a sound first reaching the nearer ear and then reaching the farther ear. This tells us what direction the sound comes from.

Sounds from anywhere inside the "cone of confusion" produce identical neural responses, so can't be told apart

Sounds outside the cone produce unique neural responses, so are easier to locate

FINDING THE SOURCE

SOUND SOURCE

Cone of confusion
In a cone-shaped region outside each ear, signals are ambiguous and we find it difficult to localize sounds. Tilting or swivelling our heads can move the sound source out of this confusing region, helping us to locate the sound.

Why does music make us emotional?

Music can cause strong emotional reactions – whether it's the soundtrack heightening fear in a scary movie, or chills created by a haunting melody. We know there are a wide range of brain areas involved in the emotions elicited, but we don't know why, or how, music creates such dramatic feelings in the listener, or why the same song affects people differently.

YOUR BRAIN ON MUSIC

WHY DO WE STAND STILL TO LISTEN?

It is easier to listen carefully when we stop moving altogether, as this helps us hear better by stopping sounds generated by our own movements.

Balancing act

As well as hearing, our ears are responsible for keeping our balance and telling us how and in which direction we are moving. They do this using a set of organs in the inner ear – one on each side of the head.

Turning and movement

Inside our ears, three canals sit at roughly 90 degrees to each other. One responds to motions such as forward rolls, the second to cartwheels, and the third to pirouettes. The relative motion of the fluid tells our brains in what direction we are moving. When spinning repeatedly in the same direction, the fluid builds up momentum. Once that matches the rate of spin, it stops deflecting the hair cells and you no longer feel motion. After stopping, however, the liquid continues, giving you the sensation you are still moving, a feeling known as dizziness.

WHY DOES ALCOHOL MAKE YOUR HEAD SPIN?

Alcohol builds up quickly in the cupulas of the inner ear and makes them float in their canals. When you lie down, the cupulas are disturbed and the brain thinks you are spinning.

This canal detects motion such as that experienced when performing cartwheels

SEMICIRCULAR CANAL

At the end of each canal is a region called an ampulla containing the sensitive hair cells

SEMICIRCULAR CANAL

AMPULLA

This canal detects forwards and backwards movements

SEMICIRCULAR CANAL

AMPULLA

This canal detects spinning or rotating motions of the head

AMPULLA

Turning sense organs

When you move, the liquid inside the canals moves too, but because it has inertia, it takes a while to start moving. This displaces a gelatinous mass called the cupula, disturbing the hair cells inside it, and sending signals to the brain. When the cupula is bent in one direction, the nerves increase their rate of firing. If it is bent in the other direction, firing is inhibited – this tells the brain the direction of the motion.

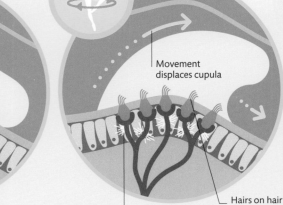

Gelatinous material

CUPULA

Movement displaces cupula

Hair cell

RESTING

Signal sent to brain

TURNING

Hairs on hair cells deflected

Steady gaze

Your brain constantly adjusts the tiny movements your muscles make to keep you balanced. Inputs from the eyes and muscles combine with those from your inner ear to determine which way up you are.

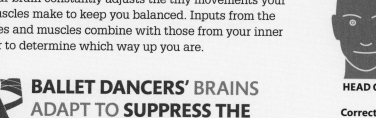

BALLET DANCERS' BRAINS ADAPT TO SUPPRESS THE SENSATION OF DIZZINESS AFTER SPINNING.

HEAD ON **TURNING RIGHT** **TURNING LEFT**

Correction reflex
Our eyes automatically correct for head movements, keeping the image on our retina stationary. Without this reflex, we would be unable to read, as the words would jump about every time our head moved.

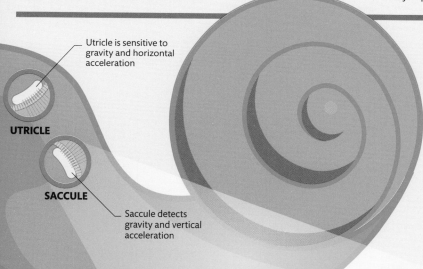

Utricle is sensitive to gravity and horizontal acceleration

UTRICLE

SACCULE

Saccule detects gravity and vertical acceleration

Gravity and acceleration

As well as turning motions, our inner ears sense straight-line acceleration – backwards and forwards, or up and down. We have two organs to sense acceleration – the utricle is sensitive to horizontal movements while the saccule detects vertical acceleration (such as the movement of a lift). Both organs also sense the direction of gravity relative to the head, such as when the head is tilted or level.

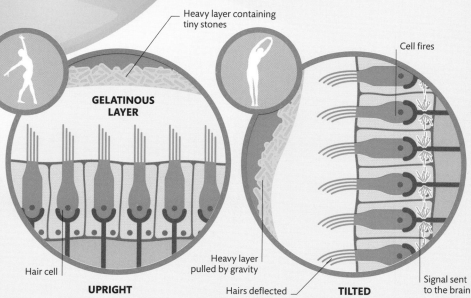

Heavy layer containing tiny stones

Cell fires

GELATINOUS LAYER

Gravity sense organs
The hair cells in the utricle and saccule are within a gelatinous layer, topped with a structure containing tiny stones. Due to the weight of the structure, gravity moves it when the head is tilted, which in turn deflects the hairs. During acceleration, the stone-filled layer takes longer to start moving because of its greater mass. If there are no other cues, it can be hard to tell the difference between a head tilt and acceleration.

Hair cell

UPRIGHT

Heavy layer pulled by gravity

Hairs deflected

TILTED

Signal sent to the brain

Hearing problems

Deafness or hearing problems are common but often treatable thanks to technological advances. Most people develop some form of hearing loss as they age due to damage to the components of the inner ear.

Causes of hearing problems

Deafness from birth is usually caused by genetic mutations that stop the ear from working properly. The hearing problems shown here can occur as a result of injury or illness throughout life.

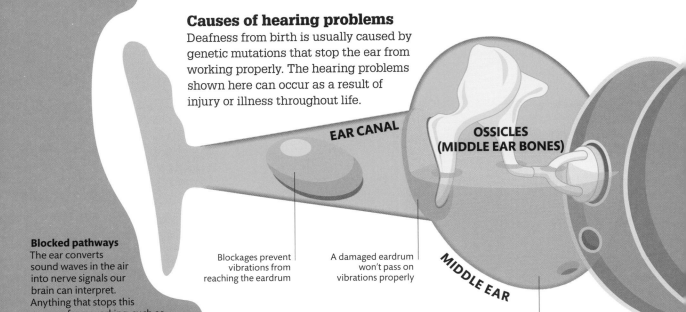

EAR CANAL

OSSICLES (MIDDLE EAR BONES)

MIDDLE EAR

Blocked pathways
The ear converts sound waves in the air into nerve signals our brain can interpret. Anything that stops this process from working, such as a physical blockage or damage, can cause hearing problems.

Blockages prevent vibrations from reaching the eardrum

A damaged eardrum won't pass on vibrations properly

Infections can cause fluid build-up, and make sounds seem muffled

How loud is too loud?

The decibel sound scale is logarithmic, and every 6 dB increase in volume doubles the sound energy. Loud noises can damage hair cells and above a certain level of damage the cells can't repair themselves, and die. If enough hair cells die, you can lose the ability to detect certain frequencies.

Causing damage
Any noise level above 85 dB can cause damage, depending on how long you are exposed to it.

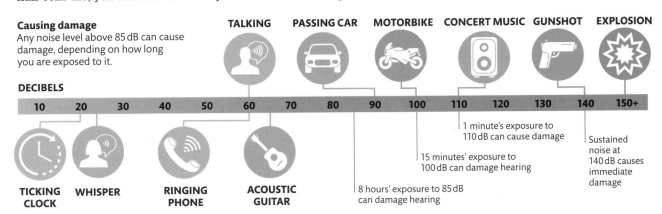

TALKING · PASSING CAR · MOTORBIKE · CONCERT MUSIC · GUNSHOT · EXPLOSION

DECIBELS

10 20 30 40 50 60 70 80 90 100 110 120 130 140 150+

1 minute's exposure to 110 dB can cause damage

15 minutes' exposure to 100 dB can damage hearing

8 hours' exposure to 85 dB can damage hearing

Sustained noise at 140 dB causes immediate damage

TICKING CLOCK · **WHISPER** · **RINGING PHONE** · **ACOUSTIC GUITAR**

AROUND AGE 18, YOU BEGIN LOSING THE ABILITY TO HEAR VERY HIGH PITCHED NOISES

BRAIN

Auditory cortex damage can cause deafness even if the ear is undamaged

NERVE

COCHLEA

Damage to the auditory nerve prevents signals reaching the brain

If hair cells are permanently damaged, certain frequencies may no longer be audible

HAIR CELLS IN THE COCHLEA

Healthy hair cells have long hairs

WHY DO LOUD NOISES MAKE YOUR EARS RING?

Loud noises vibrate hair cells so violently that the tips can snap off, causing them to send signals to your brain after the noise has finished. The tips can grow back within 24 hours.

Cochlear implants

Normal hearing aids simply amplify sounds and cannot help people with damaged or missing hair cells. Cochlear implants replace the function of the hair cells, converting sound vibrations into nerve signals that the brain learns to interpret. More current through the electrodes within the cochlea produces a louder sound, while the position of the activated electrodes determines pitch.

How they work

External microphones detect sounds and send them to the processor. Signals then travel to the internal receiver via the transmitter, before passing as electrical current to the electrode array inside the cochlea. Stimulated nerve endings send signals to the brain, and sounds are heard.

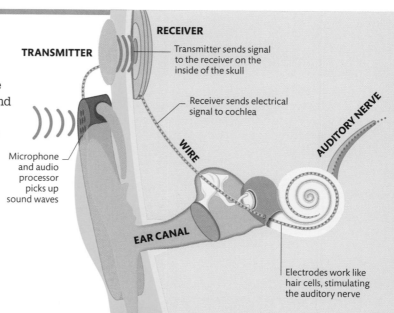

TRANSMITTER

RECEIVER

Transmitter sends signal to the receiver on the inside of the skull

Receiver sends electrical signal to cochlea

AUDITORY NERVE

WIRE

Microphone and audio processor picks up sound waves

EAR CANAL

Electrodes work like hair cells, stimulating the auditory nerve

Catching a scent

Particles in the air are detected by sensory cells in your nose, and signals are sent to your brain so you can identify them as smells. Smells can invoke powerful emotions or memories because of physical links to your brain's emotional centre.

Sense of smell

Anything that smells releases tiny particles, or scent molecules, into the air. When you inhale, these molecules pass into your nose, where the smell is detected by specialized nerve cells. Sniffing is an automatic response when catching a whiff – the more scent molecules you inhale, the easier it is to identify a smell. Our senses of smell and taste often work hand-in-hand when enjoying a meal, because scent molecules are released by the food we eat, which then pass into the rear of the nasal cavity.

HUMANS HAVE AROUND 12 MILLION RECEPTOR CELLS AND THEY CAN DETECT 10,000 DIFFERENT ODOURS!

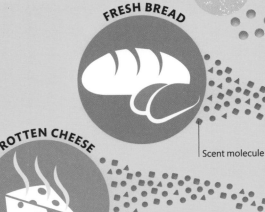

2 Nose hair
At the entrance to the nose, hairs catch large particles of dust and debris, but admit the scent molecules, which are millions of times smaller.

DUST

FRESH BREAD

ROTTEN CHEESE

Scent molecule

SMOKE

1 Types of smell
Aromatic objects such as freshly baked bread, off cheese, and burning things, release scent molecules. The type of molecules determines what you smell as well as the smell's intensity, since we are much more sensitive to some scent molecules than others.

LOSS OF SMELL

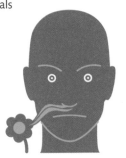

A complete lack of smell is called anosmia. Some people are born with anosmia, while others develop the condition after an infection or head injury. These instances can cause a severage of the nerve fibres, reducing the number of nervous signals they pass to the brain. Those with anosmia have reduced appetites and are more likely to suffer from depression – this is probably because of smell's links with the brain's emotional centre. The sense can recover on its own or after drug treatment or surgery. For others, smell training, which probably leads to the regeneration of olfactory receptor cells, can help.

WHY DO WE HAVE NOSE BLEEDS?

Nasal membranes that line your nasal cavity are thin and filled with tiny blood vessels. These blood vessels can burst very easily to cause nose bleeds – either by breathing dry air which crusts and breaks the thin membrane or even by blowing your nose too hard.

3 Nasal cavity
Scent molecules waft into the nasal cavity as we breathe in. Specialized nerve cells, called olfactory receptors, sit at the top of each cavity and detect scent molecules. Thin, bony conchae radiate warmth to keep the olfactory receptors functioning and healthy.

Olfactory bulb full of nerves carrying smell signals to brain

PLEASURE

DISGUST

FEAR

AMYGDALA

5 Smell and emotion
The smell of fresh food often inspires pleasure. Smelling anything "off" will cause disgust and alert you to a risk of illness, and the scent of smoke can kickstart the fight-or-flight response.

OLFACTORY RECEPTORS

NERVES

Conchae full of blood vessels warm air

4 To the brain
Nerve signals are sent from the tips of the olfactory receptors to nerve fibres packed inside the olfactory bulb. Signals then travel to the amygdala, where the emotional reaction to each smell is established.

Nose hair catches dust and harmful bacteria

Incoming air warmed by blood vessels in nose

Lock and key theory
Each of your olfactory receptors responds to particular groups of scent molecules, just as certain keys fit into certain locks. Different patterns of receptors are activated by different smells, therefore we can identify more smells than we have receptors. Whether it is the shape of the molecule that determines where it binds or a different factor entirely is under debate.

Mucus-secreting gland

Olfactory receptor cell

Supporting cell

Olfactory receptor cell may receive two types of scent molecules

First type of scent molecule

Second type of scent molecule

One type of receptor for one type of scent molecule

Mucus

Scent molecule dissolving in mucus

Olfactory receptors
Scent molecules in the nasal cavity dissolve into a thin layer of mucus. This allows the molecules to bind to the ends of the olfactory receptor cells.

On the tip of the tongue

Your tongue has thousands of chemical receptors, which detect some key chemical ingredients in your food and interpret them as one of five major taste sensations. However, not everyone's tongue is the same, which helps explain food preferences.

Taste receptors

Our tongues are covered in tiny bumps (papillae), which contain taste receptors for chemicals that give us the five basic tastes – sour, bitter, salty, sweet, and umami (savoury). Each receptor deals only with one taste, and there are receptors for all five tastes all over the tongue's surface. The flavour of food is a more complex sensation, comprised of taste mixed with smell, detected when molecules travel up the back of the throat into the nose. This is why things taste bland when your nose is blocked.

SOUR

A papilla – a visible bump on the tongue that may contain taste buds sensitive to sour, bitter, salty, sweet, or umami tastes

BITTER

SALTY

Taste bud

A taste bud begins with a pore on the tongue papilla's surface. The pore lets in particles of food or drink, which contact taste receptor cells. The cells send signals to the brain when certain tastes are detected. Taste buds are also found on the insides of your mouth.

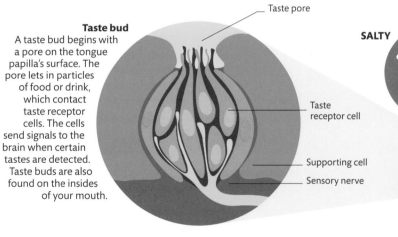

Taste pore

Taste receptor cell

Supporting cell

Sensory nerve

UMAMI

SWEET

SUPERTASTERS

Some people have many more taste buds than others. These supertasters can detect bitter substances that other people can't and generally dislike green vegetables and fatty foods. Supertasters are thought to make up 25 per cent of the population.

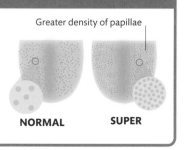

Greater density of papillae

NORMAL

SUPER

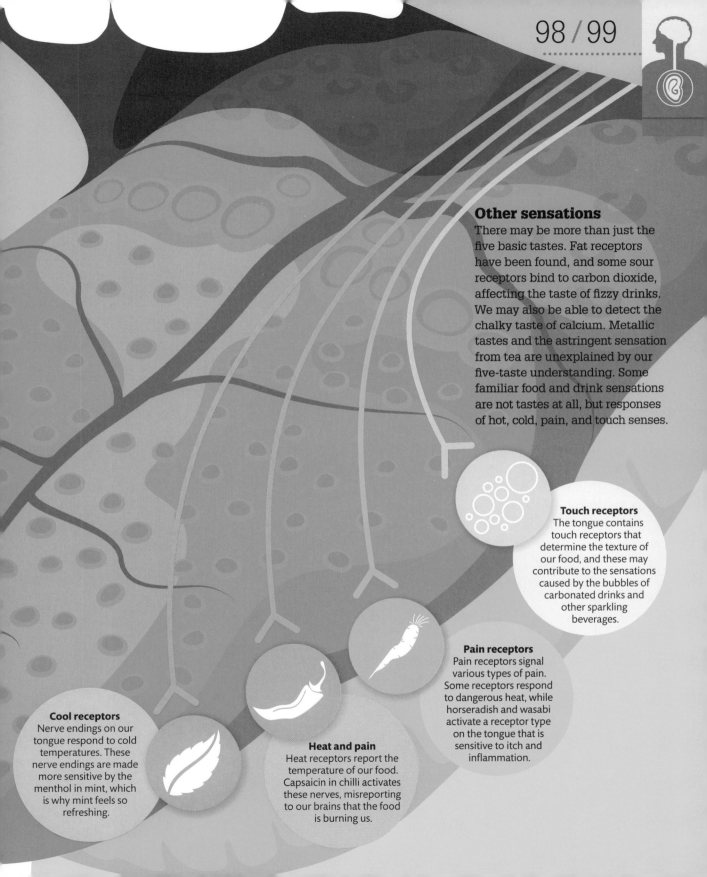

Other sensations

There may be more than just the five basic tastes. Fat receptors have been found, and some sour receptors bind to carbon dioxide, affecting the taste of fizzy drinks. We may also be able to detect the chalky taste of calcium. Metallic tastes and the astringent sensation from tea are unexplained by our five-taste understanding. Some familiar food and drink sensations are not tastes at all, but responses of hot, cold, pain, and touch senses.

Touch receptors

The tongue contains touch receptors that determine the texture of our food, and these may contribute to the sensations caused by the bubbles of carbonated drinks and other sparkling beverages.

Pain receptors

Pain receptors signal various types of pain. Some receptors respond to dangerous heat, while horseradish and wasabi activate a receptor type on the tongue that is sensitive to itch and inflammation.

Heat and pain

Heat receptors report the temperature of our food. Capsaicin in chilli activates these nerves, misreporting to our brains that the food is burning us.

Cool receptors

Nerve endings on our tongue respond to cold temperatures. These nerve endings are made more sensitive by the menthol in mint, which is why mint feels so refreshing.

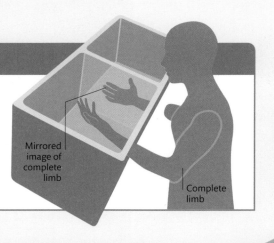

Mirrored image of complete limb

Complete limb

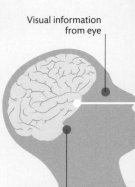

Visual information from eye

Balance information from ear

Body position sense

How do you know where your hand is if you're not looking at it? Sometimes called our sixth sense, we have receptors dedicated to telling our brains where each part of our body is in space. We also get a sense that our body parts belong to us.

Tension receptor
Organs within your tendons detect how much force your muscles are exerting by monitoring muscular tension (see pp.56–57).

Muscle

Golgi tendon organ senses changes in muscle tension

Tendon

Bone

Position sensors

There is a range of different receptors that help the brain calculate the position of our body. For a limb to move, the joint must change position. Muscles either side of the joint contract or relax, changing in length or tension. Tendons that attach muscle to bone are stretched, as is the skin on one side of the joint, while the skin on the other side relaxes. By combining information about each of these components, the brain can construct a fairly accurate picture of the body's movements.

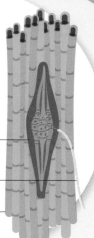

Stretch receptor
Tiny spindle-shaped sense organs buried in your muscles detect changes in the length of the muscle, telling the brain how contracted the muscle is.

Muscle spindle organ detects changes in length of muscle

Nerve sends signal to brain

Muscle

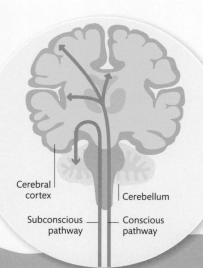

Cerebral cortex

Cerebellum

Subconscious pathway

Conscious pathway

Integrator

The brain combines information from the sensors located in and around the muscles as well as your other senses to interpret how your body is positioned. The conscious element of this is controlled by the cerebral cortex and allows you to run, dance, or catch. The cerebellum, at the base of the brain, is in charge of the unconscious elements that keep you upright without you thinking about it.

Bone

Touch sensitive nerves

Joint receptors
Receptors within your joints detect its position. They are most active when your joints are at their extremes to help prevent damage through over-extension. However, they may also play some role in detecting the position of joints in normal motion.

Ligament receptors

Ligament

BODY OWNERSHIP SENSE

Your sense that your body is your own is more complicated and flexible than it seems. The rubber hand illusion induces the feeling that a fake hand belongs to you. A similar technique can invoke out-of-body experiences, using a virtual reality headset. This flexibility allows us to cope if we lose a limb, or to include tools and prosthetics in what we think of as "part" of our body.

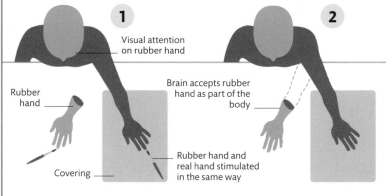

1 Visual attention on rubber hand

Rubber hand

Covering

2 Brain accepts rubber hand as part of the body

Rubber hand and real hand stimulated in the same way

ESTABLISHING THE CONNECTION

RUBBER HAND SEEN AS PART OF BODY

Skin stretch
Special receptors in the skin (see p.75) can detect stretch. This helps us determine the movements of our limbs, particularly changes in the angle of a joint, which causes the skin on one side to stretch while the skin on the opposite side is slackened.

BODY POSITION SENSORS
IN THE JAW MUSCLES
AND TONGUE HELP
YOU **FORM THE
RIGHT SOUNDS**
WHEN YOU SPEAK

Integrated senses

Your brain makes sense of the world around you by combining information from all your senses. But, surprisingly, sometimes one sense can actually change how you experience another.

How senses can interact

Everything you experience is interpreted by your senses. When you see and pick up an item, you feel its shape and texture. You look for where sounds or smells are coming from and "eat with your eyes" before tasting your food. Your brain performs complex processing to integrate this information correctly. Sometimes, this combination of information can cause multi-sensory illusions. If information from different senses seems to conflict, the brain favours one sense over another, and depending on the situation, this can be helpful or misleading.

BRAIN

Sound and vision
When things happen simultaneously you often assume they are linked. For example, if you hear an alarm close enough to where your car is parked you will disregard the location of the sound and believe the alarm is coming from your car.

Sound of alarm can be distinguished from car

CAR ALARM SOUND

Sound of car alarm is close to car

CAR

You move towards car assuming it is the alarm's source

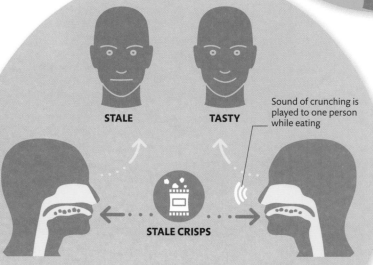

STALE   **TASTY**

Sound of crunching is played to one person while eating

STALE CRISPS

Taste and sound
If someone listens to the sound of crunching while eating stale crisps, they will claim they taste fresh. Tactically, manufacturers make crisp bags crackly so that their crisps seem crunchier.

IN NOISY ENVIRONMENTS YOU **LIP-READ**, USING **WHAT YOU SEE** TO INTERPRET **MUFFLED SPEECH**

SOUNDS AND SHAPES

When shown these shapes and asked to name one Bouba and the other Kiki, most people call the spiky shape Kiki because of its spiky sound, while deciding the softer Bouba fits the rounded shape. This pairing holds across a wide variety of cultures and languages, indicating a link between the senses of sound and sight.

Smell and taste
Taste is a simple sense, made up of crude sensations such as "sweet" or "salty". Most of what you think of as flavour is actually what you are smelling. Smell can also influence the crude sense of taste itself. Smelling vanilla can make food or drink taste sweeter, but only in parts of the world where vanilla is a common flavour for sweet foods.

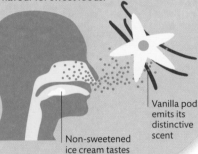

Vanilla pod emits its distinctive scent

Non-sweetened ice cream tastes sweet

Image of ball and spring bouncing on virtual version of hand

Pressure of ball and spring felt on real hand

VIRTUAL REALITY

REAL LIFE

Touch and vision
When gamers pick up objects in virtual reality, visual cues give them physical sensations, even though their touch sense gives them no such information. What your eyes can see can actually influence what you feel.

Using your voice

Talking is achieved by a complex yet flexible network of neural pathways in the brain and physical coordination of the body. Tone and inflection influences how words are spoken, which can add numerous meanings to even the simplest of sentences.

1 **Thought process**
Firstly, you must decide what words you want to say. This activates a network of regions in the left hemisphere of your brain, including Broca's area, drawing on your memory store of words.

Broca's area on the left side of brain formulates speech

3 **Producing sound**
As you exhale, your vocal cords vibrate as the air stream passes them, making sound. Vibration speed dictates your voice's pitch, and this is controlled by muscles in the larynx. If you want to shout, you need a stronger airstream.

Vibrating cords cause sound

Vocal cords open to allow air into lungs

Larynx

2 **Breathing in**
Your lungs provide the constant stream of air that you need in order to speak. When inhaling, vocal cords open to allow air to pass through, and then air pressure begins to build in the lungs.

4 **Articulation**
Your nose, throat, and mouth act as resonators, while lip and tongue movements introduce specific sounds, altering the buzzing produced by the vocal cords into recognisable speech.

Air pressure in lungs builds

MAKING AN "AA" SOUND

MAKING AN "EE" SOUND

MAKING AN "OO" SOUND

How do you talk?

The brain, lungs, mouth, and nose all play vital roles when producing speech, but the voicebox, or the larynx, is the most important. Located in your throat above your windpipe, it contains two sheets of membrane that stretch across the inside. These are the vocal cords, and they are the structures that produce the sound you craft into speech.

Making different sounds
Your tongue moves to mould sounds created by your vocal cords, aided by the teeth and lips. Changing the shape of your tongue and mouth produces vowels like "aah", or "eee", and the lips interrupt air flow to produce consonants, such as "p" and "b".

Pathway of speech
Each area of the brain is connected via nerves. The bundle of nerves linking Wernicke's and Broca's areas, the arcuate fasciculus, is comprised of nerve cells that fire at high speeds.

MOTOR CORTEX

Motor cortex sends instructions to muscles to articulate reply

BROCA'S AREA

Bundle of nerves links Wernicke's and Broca's areas

Broca's area allows listener to plan reply based on speech heard

AUDITORY AREA

WERNICKE'S AREA

Auditory area analyzes speech

Wernicke's area processes word meanings

Speech reaching listener's ear

Processing speech

Air vibrations caused by speech reach the ear and trigger nerve cells deep inside, which then send signals to the brain for processing. Wernicke's area is vital for understanding the basic meaning of the words, while Broca's area interprets grammar and tone. These regions are part of a larger network which understands and produces speech. Damage to either area can lead to speech problems.

HOW DO YOU SING?

When you sing, you use the same physical and cognitive networks as when you speak, but it requires much more control. Air pressure is greater, and several chambers, such as the sinuses, mouth, nose, and throat are used as resonators, producing a richer sound.

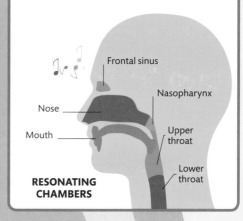

Frontal sinus

Nasopharynx

Nose

Mouth

Upper throat

Lower throat

RESONATING CHAMBERS

Reading faces

We are a social species, so recognizing and understanding faces is vital for our survival. This means we have evolved to be very good at noticing them – even sometimes seeing them where they don't really exist, like on a piece of burnt toast!

Importance of understanding faces

From birth, babies are fascinated by faces, and show a preference to looking at them above everything else. As you age, you not only quickly become an expert in recognizing faces, but also reading expressions. This allows you to identify those who would help or harm you. Individual faces can stay in your memory for a remarkable length of time, even if you haven't seen the person in years.

Facial expression cues
When recognizing a face, you look at the ratio between the eyes, nose, and mouth. Movements of these can help you detect emotions; for example, raised eyebrows and an open mouth would signal surprise. These signals are interpreted by your eyes and nervous signals are sent to the fusiform face area in your brain to be processed.

Fusiform face area
This area of the brain, named the fusiform face area, is activated when you look at faces. It is thought that this area of the brain is specialized in facial recognition. However, it also becomes active when you are looking at objects with which you are familiar – if you were a pianist, it may become active when you see a keyboard. Whether it is face-specific is still under debate.

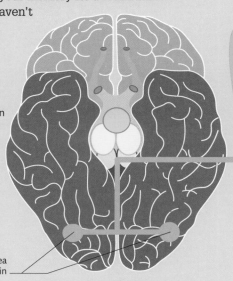

Location of fusiform face area on both sides of brain

UNDERSIDE OF BRAIN

RECOGNIZING FACES

Humans tend to spot faces in random patterns and places – from cars to grilled cheese sandwiches to pieces of wood. This is because it was better for our ancestors to interpret the faces of others in order to thrive in a complex social hierarchy.

Expression muscles

Your face contains muscles that pull your skin and change the shape of your eyes and position of your lips, making your face highly expressive. The ability to read these expressions on other faces allows you to judge other people's moods, intentions, and meanings. Faces tell us when to ask for a favour, when to leave a person alone, or when to offer comfort. Picking up even the subtlest cues, such as the furrowing of the brow or the curling of the lip, can mean the difference between interpreting a frown or a smirk correctly.

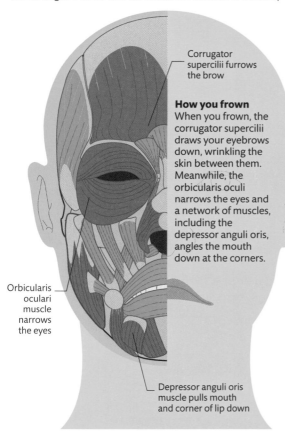

Corrugator supercilii furrows the brow

How you frown
When you frown, the corrugator supercilii draws your eyebrows down, wrinkling the skin between them. Meanwhile, the orbicularis oculi narrows the eyes and a network of muscles, including the depressor anguli oris, angles the mouth down at the corners.

Orbicularis oculari muscle narrows the eyes

Depressor anguli oris muscle pulls mouth and corner of lip down

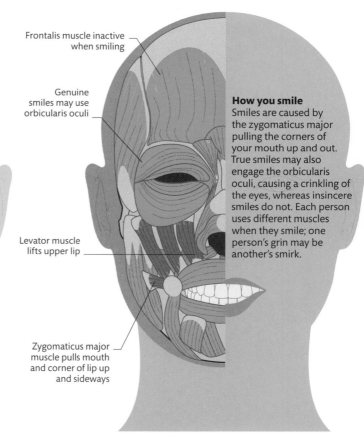

Frontalis muscle inactive when smiling

Genuine smiles may use orbicularis oculi

How you smile
Smiles are caused by the zygomaticus major pulling the corners of your mouth up and out. True smiles may also engage the orbicularis oculi, causing a crinkling of the eyes, whereas insincere smiles do not. Each person uses different muscles when they smile; one person's grin may be another's smirk.

Levator muscle lifts upper lip

Zygomaticus major muscle pulls mouth and corner of lip up and sideways

GAZE AND EYE CONTACT

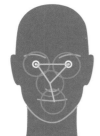

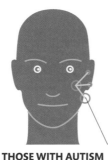

People with autism (see p.246) usually don't focus on the eyes and mouth when looking at faces. They find socializing confusing and difficult, and may miss vital social cues when communicating. Babies may even exhibit this averted gaze, and they may go on to develop the condition, so it could be used as an early warning sign for autism.

People with autism show different patterns of looking behaviour

TYPICAL GAZE　　**THOSE WITH AUTISM**

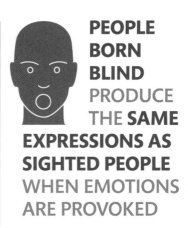

PEOPLE BORN BLIND PRODUCE THE **SAME EXPRESSIONS AS SIGHTED PEOPLE** WHEN EMOTIONS ARE PROVOKED

What you don't say

You communicate using more than just your words. Facial expressions, tone of voice, and hand gestures can speak volumes, and noticing these signals is vital for understanding what someone really means.

Non-verbal communication

When you are talking to someone, you are subconsciously picking up on subtle signals from their voice, face, and body. Interpreting these signals correctly is most important when what is said could be ambiguous. Most of these signals allow you to gauge the mood of a person or group so you act appropriately in social situations. For example, in a meeting at work, assessing the body language and moods of your colleagues can be advantageous to you if you are waiting for the right time to pitch a big idea.

INVADING SOMEONE'S PERSONAL SPACE CAN INSPIRE **FEAR, AROUSAL, OR DISCOMFORT**

Types of signals
Facial expressions, hand gestures, body posture, and the tone and speed of somebody's voice are all signals you process when communicating. What someone is wearing is also important, as it can provide clues about their personality, religion, or culture. Physical contact can add emotional weight to what is said.

FACIAL EXPRESSIONS

TYPE OF CLOTHING

HAND GESTURES

BODY POSTURE

TONE AND SPEED OF VOICE

PHYSICAL CONTACT

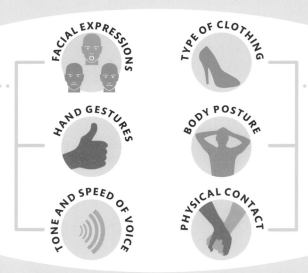

Arms folded, forming barrier

Body turned away from others

Head tilted

Physical contact

Mirrored legs

NEGATIVE   **POSITIVE**

Body language
The way your body moves as you speak can often be just as telling as what you say. Holding eye contact, mirroring the facial expressions and posture of others, and physical contact, are generally interpreted as positive signals. Folded arms, hunched shoulders, and positioning yourself away from others can produce negative vibes.

Micro-expression

1 SECOND

Pausing
You tend to pause more when you lie, as thinking of a fabricated response takes longer than providing a natural one. Even if you are telling a story that happened, and it is only your emotions towards the event that are untrue, pausing is still a tell-tale sign of lying.

Caught in a lie
It is sometimes an advantage to deceive those around you, but also useful to be able to tell when someone is deceiving you. However, there are signals that can give you away when you lie. The best liars convince themselves they are telling the truth – if you truly believe your lie, your body language can't give you away.

Micro-expressions
Lightning-quick expressions appear unconsciously on the face of a liar and usually show an emotion he or she is attempting to conceal. They last under half a second and are usually missed by the average person, but can be detected by a trained viewer.

Visible hand twitches can be a giveaway

Hand movements
Movements of the body are unedited by consciousness so are often a more reliable indicator of lying. When you lie, you often wring your hands, make gestures, or have nervous twitches.

IS IT POSSIBLE TO DETECT ALL LIES?

No, everyone has different ticks and ways of lying. One person may pause and one may twitch their toes, so there is no foolproof system to detect all lies.

Twitches of a person's toes can be an indicator of lies

SUPERMAN POSE

Body language is so powerful that it can even change the way you feel about yourself. Adopting a powerful stance for just 1 minute raises your levels of testosterone, in both men and women, and reduces levels of the stress hormone cortisol. This increases feelings of control, the likelihood that you will take risks, and your performance in job interviews improves too. This shows that movements of your body can influence emotions, and proves the old saying "fake it till you make it" really is good advice!

THE HEART
OF THE
MATTER

Filling your lungs

Your lungs act like a giant pair of bellows, drawing air in and letting it out to extract oxygen and expel waste carbon dioxide. You breathe around 12 times per minute at rest and 20 times per minute or more during exercise; which all adds up to roughly 8.5 million breaths per year.

Controlling breaths

Your breathing rate speeds up or slows down due to signals from chemical receptors in the blood vessels. These receptors provide a feedback loop between the blood vessels, brain, and diaphragm.

Drawing breath

Air drawn in through the nose or mouth passes down the trachea, or windpipe, which channels air into the left or right bronchus, and then into smaller and smaller air passages called bronchioles. Between the trachea and the ends of the bronchioles, your airway divides 23 times.

Feedback system

Chemical receptors detect changes in oxygen, carbon dioxide, and acidity levels in the blood. This information is sent to the brain, which controls the diaphragm's movements, increasing or decreasing rate and depth of breathing to keep blood levels constant.

1 **Breathing in**
Air is warmed and moistened as it passes through the nose or mouth. Nasal hairs filter out dust particles that could irritate the trachea or lungs and cause a coughing fit.

TRACHEA

NASAL CAVITY

TONGUE

Air breathed in

Air travelling through the throat

Air travelling down the trachea

LUNG

Bronchiole

Lining of right lung

SIGNAL TO BRAIN

BRAIN

HEART

DIAPHRAGM

NERVE

Blood vessel

Cluster of receptors monitors levels of oxygen in blood from the heart

Receptor monitors levels of oxygen in blood vessels

Direction of nerve signals

Signals sent to diaphragm to control breathing rate

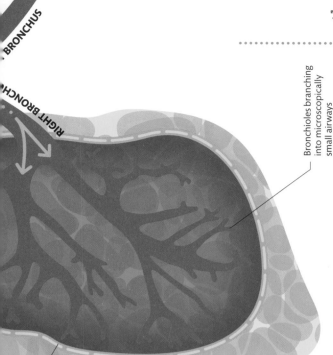

BRONCHUS

RIGHT BRONCHUS

Bronchioles branching into microscopically small airways

Pleural cavity

2 Into the lungs

Air travels down each bronchus into ever-smaller passages, eventually ending in tiny air sacs called alveoli. The lungs are separated from the chest by a pleural cavity filled with pleural fluid. This thin layer of fluid acts as a sticky lubricant, letting your lungs slide over your chest wall and preventing them from pulling away as you breathe out.

ALL YOUR AIRWAYS LAID END TO END WOULD MEASURE **2,400 KM** (1,490 MILES)

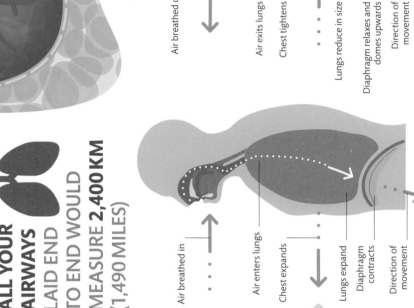

EXHALING

Air breathed out

Air exits lungs

Chest tightens

Lungs reduce in size

Diaphragm relaxes and domes upwards

Direction of movement

INHALING

Air breathed in

Air enters lungs

Chest expands

Lungs expand

Diaphragm contracts

Direction of movement

SIZE MATTERS

The surface area of all the tiny air sacs (alveoli) in the lungs measures an incredible 70 sq m (750 sq ft) – this is 40 times greater than the surface area of your skin! This maximizes the amount of oxygen you can absorb.

SKIN

LUNGS

Mechanics of breathing

Chest muscles and the rib cage influence breathing, but the main powerhouse is the diaphragm. It is a large domed muscle that separates the chest from the lower organs. To breathe in, the diaphragm contracts and pulls down like a piston. At the same time, muscles between your ribs contract, lifting your ribs so your lungs expand and air rushes in. When your diaphragm and chest muscles relax, air is forced out.

From air to blood

Every cell in your body needs oxygen and your lungs are highly adapted to extract this life-sustaining gas from the atmosphere. This extraction occurs from 300 million tiny air sacs called alveoli, which give your lungs a spongelike texture.

Deeper into the lungs

Inhaled air passes from the throat into the trachea to reach tiny branches called bronchioles. Mucus covers each bronchiole, which keeps them moist and traps inhaled particles. Each bronchiole is lined with thin strips of muscle. In people with asthma, a sudden constriction of these muscles narrows the airways, causing shortness of breath.

THE AIR YOU BREATHE OUT CONTAINS 16 PER CENT OXYGEN, ENOUGH TO RESUSCITATE SOMEONE!

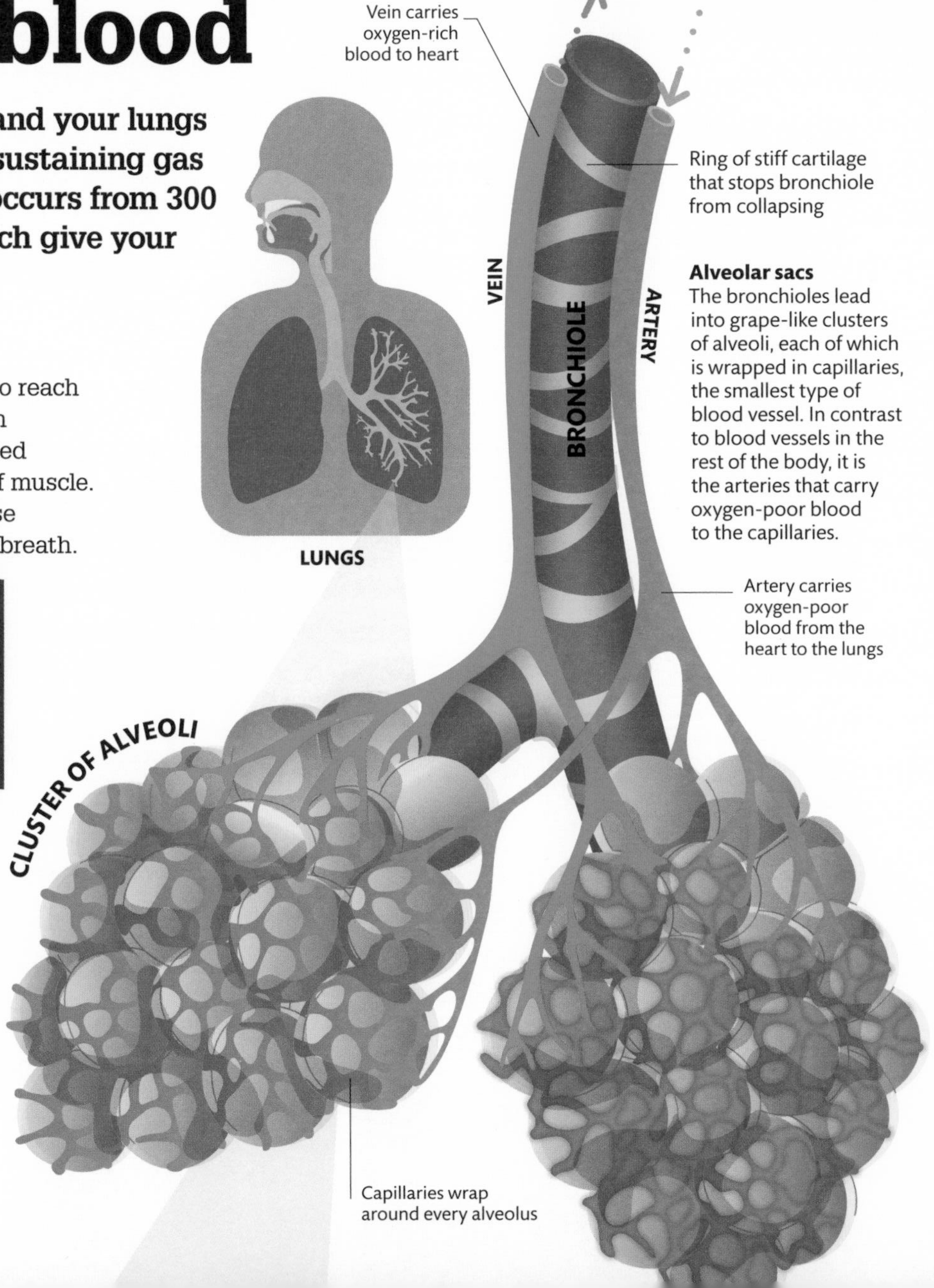

WHY CAN WE SEE OUR BREATH IN COLD AIR?

The air you breathe is warmed in your lungs, so when you exhale, water vapour in your breath condenses into clouds of water droplets.

Ring of stiff cartilage that stops bronchiole from collapsing

Alveolar sacs
The bronchioles lead into grape-like clusters of alveoli, each of which is wrapped in capillaries, the smallest type of blood vessel. In contrast to blood vessels in the rest of the body, it is the arteries that carry oxygen-poor blood to the capillaries.

Artery carries oxygen-poor blood from the heart to the lungs

Vein carries oxygen-rich blood to heart

ARTERY

BRONCHIOLE

VEIN

LUNGS

CLUSTER OF ALVEOLI

Capillaries wrap around every alveolus

HIGH ALTITUDE

At high altitudes, air is thinner and less oxygen is present. You may find yourself automatically taking deep breaths, since your body will detect lower amounts of oxygen in your bloodstream than it normally expects.

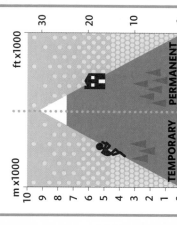

m x1000		ft x1000
10		30
9		
8		
7		20
6		
5		
4		10
3		
2		
1		
0		0

TEMPORARY   **PERMANENT**

Acclimatizing
People who travel to high altitudes can adapt by producing more red blood cells to carry more oxygen in their circulation. Full adaptation takes around 40 days, but is not permanent.

Adapting
Those who live their entire lives at high altitudes may inherit larger lungs, wider chests, and more efficient oxygen-processing genes in order to cope permanently with the hardships.

Blood going back to heart to be pumped around the body

CAPILLARY

Oxygenated red blood cell

Oxygen entering red blood cell

2   Oxygen

The oxygen we breathe diffuses from alveolar air into the blood. Here, it is captured by red blood cells, turning them, and the blood, bright red.

Exhaled air contains 100 times more carbon dioxide than inhaled air

Inhaled air contains 21 per cent oxygen

One-cell-thick wall of alveolus

ALVEOLUS

KEY
- ⋯→ Oxygen
- ⇒ Carbon dioxide

One-cell-thick wall of capillary

Blood plasma rich in carbon dioxide

Oxygen-poor red blood cell

Carbon dioxide entering air

1   Carbon dioxide

Carbon dioxide diffuses from the blood plasma through the one-celled walls of the capillary and alveolus. Blood can absorb oxygen and get rid of carbon dioxide simultaneously.

Gas exchange

Capillaries are in such close contact with alveoli that gases are able to cross over rapidly. Carbon dioxide leaves the blood in exchange for oxygen, and the newly oxygenated blood is distributed around the body by the heart. Since you do not exhale all your inhaled air in one breath, oxygen-poor and oxygen-rich air mixes in your lungs, which is why exhaled air contains some oxygen.

Why do we breathe?

The oxygen we breathe is vital for staying alive because we use it to create energy. Tiny capillaries, the smallest type of blood vessel, transport oxygen to the 50 trillion cells that make up your body.

One person uses about 550 litres (968 pints) of oxygen per day.

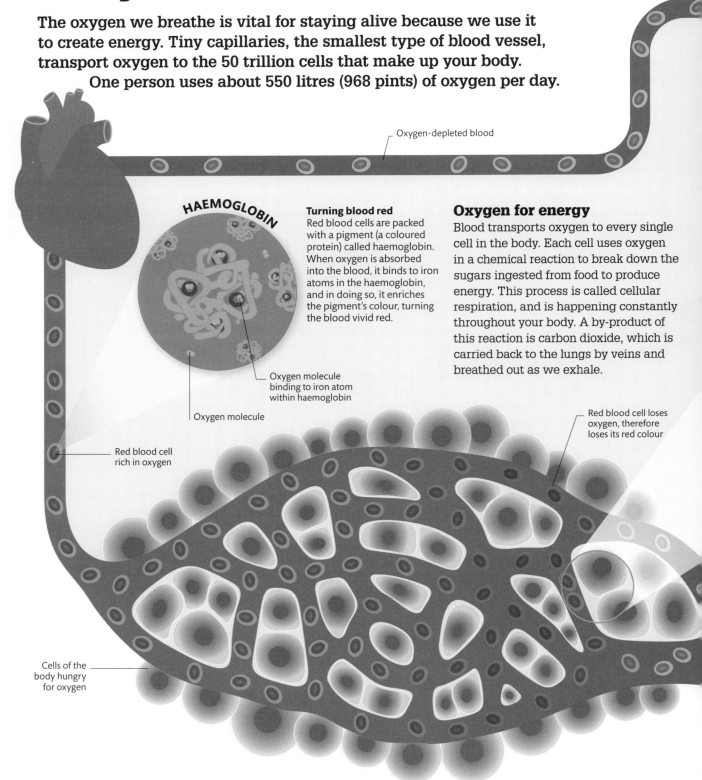

Oxygen-depleted blood

HAEMOGLOBIN

Turning blood red

Red blood cells are packed with a pigment (a coloured protein) called haemoglobin. When oxygen is absorbed into the blood, it binds to iron atoms in the haemoglobin, and in doing so, it enriches the pigment's colour, turning the blood vivid red.

Oxygen for energy

Blood transports oxygen to every single cell in the body. Each cell uses oxygen in a chemical reaction to break down the sugars ingested from food to produce energy. This process is called cellular respiration, and is happening constantly throughout your body. A by-product of this reaction is carbon dioxide, which is carried back to the lungs by veins and breathed out as we exhale.

Oxygen molecule binding to iron atom within haemoglobin

Oxygen molecule

Red blood cell rich in oxygen

Red blood cell loses oxygen, therefore loses its red colour

Cells of the body hungry for oxygen

Gas exchange

Oxygen diffuses, or drifts, from where it is in high concentrations (in red blood cells) to where there is a low concentration (in body cells). Likewise, carbon dioxide diffuses from these body cells into the blood.

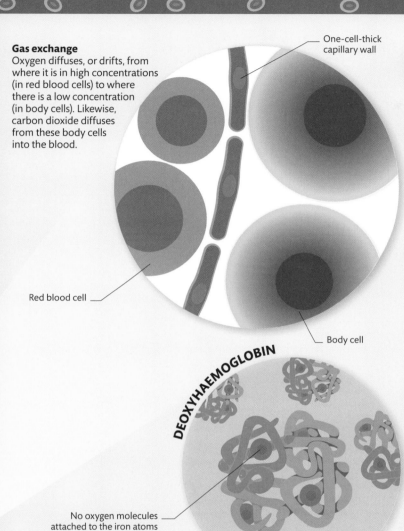

One-cell-thick capillary wall

Red blood cell

Body cell

THIN CAPILLARIES

Capillaries connect tiny arteries (arterioles) with tiny veins (venules). The thin walls of capillaries allow the exchange of oxygen and carbon dioxide. They are thin enough to access all body tissues from bones to skin, yet only just wide enough for red blood cells. Red blood cells even have to change their shape to squeeze through some capillaries.

**HUMAN HAIR
0.08MM**

**BLOOD CAPILLARY
0.008MM**

DEOXYHAEMOGLOBIN

No oxygen molecules attached to the iron atoms in the deoxyhaemoglobin

Blue blood?

When haemoglobin is carrying oxygen, it is called oxyhaemoglobin. When it releases oxygen into your body tissues, it becomes deoxyhaemoglobin, and turns a dark red colour – the colour of oxygen-depleted blood. The blood is not really blue, even though veins look blue beneath your skin.

IF YOU **HOLD YOUR BREATH,** THERE IS ENOUGH **OXYGEN** IN YOUR **BLOOD** TO **STAY CONSCIOUS** FOR **SEVERAL MINUTES**

Red blood cell without oxygen

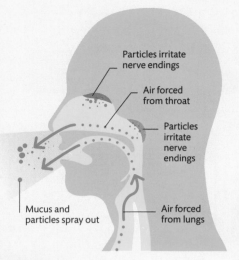

Sneezing

This reflex aims to remove irritants from the nasal cavities, and can be triggered by inhaled particles, infection, or allergies.

Particles irritate nerve endings

Air forced from throat

Particles irritate nerve endings

Mucus and particles spray out

Air forced from lungs

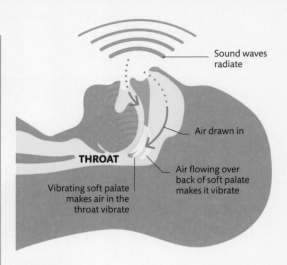

Snoring

A partial collapse of the upper airway during sleep will cause snoring. The tongue falls back and the soft palate vibrates as you breathe.

Sound waves radiate

Air drawn in

THROAT

Air flowing over back of soft palate makes it vibrate

Vibrating soft palate makes air in the throat vibrate

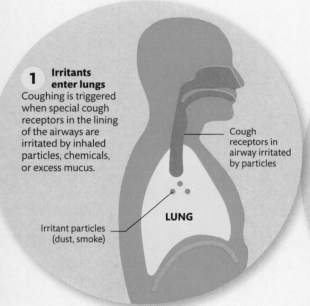

1 Irritants enter lungs
Coughing is triggered when special cough receptors in the lining of the airways are irritated by inhaled particles, chemicals, or excess mucus.

Cough receptors in airway irritated by particles

LUNG

Irritant particles (dust, smoke)

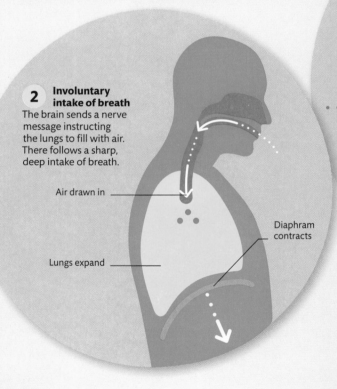

2 Involuntary intake of breath
The brain sends a nerve message instructing the lungs to fill with air. There follows a sharp, deep intake of breath.

Air drawn in

Lungs expand

Diaphram contracts

Coughs and sneezes

The respiratory system leaps into sudden action without our conscious control. Its reflex actions get rid of particles in the airways with coughs and sneezes. The functions of hiccups and yawns, however, are more mysterious.

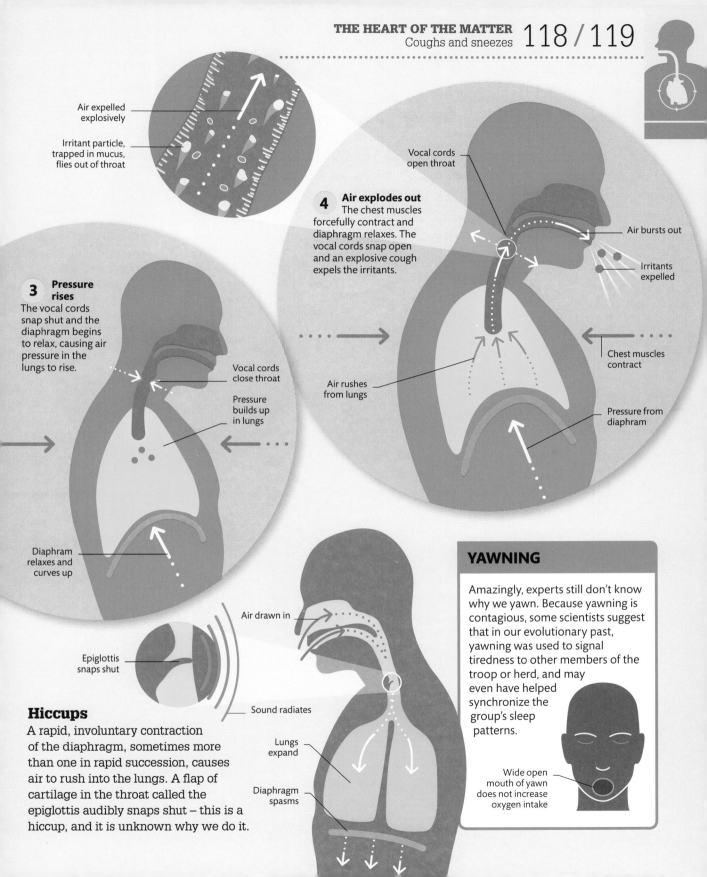

Air expelled explosively

Irritant particle, trapped in mucus, flies out of throat

4 Air explodes out
The chest muscles forcefully contract and diaphragm relaxes. The vocal cords snap open and an explosive cough expels the irritants.

Vocal cords open throat

Air bursts out

Irritants expelled

3 Pressure rises
The vocal cords snap shut and the diaphragm begins to relax, causing air pressure in the lungs to rise.

Vocal cords close throat

Pressure builds up in lungs

Air rushes from lungs

Chest muscles contract

Pressure from diaphram

Diaphram relaxes and curves up

Hiccups

A rapid, involuntary contraction of the diaphragm, sometimes more than one in rapid succession, causes air to rush into the lungs. A flap of cartilage in the throat called the epiglottis audibly snaps shut – this is a hiccup, and it is unknown why we do it.

Epiglottis snaps shut

Air drawn in

Sound radiates

Lungs expand

Diaphragm spasms

YAWNING

Amazingly, experts still don't know why we yawn. Because yawning is contagious, some scientists suggest that in our evolutionary past, yawning was used to signal tiredness to other members of the troop or herd, and may even have helped synchronize the group's sleep patterns.

Wide open mouth of yawn does not increase oxygen intake

The many tasks of your blood

Your heart and blood vessels contain around 5 litres (10½ pints) of blood, which transports everything your cells need or produce, such as oxygen, hormones, vitamins, and wastes. Blood carries the nutrients from food to the liver for processing, takes toxins to the liver for detoxification, and transports wastes and excess fluid to the kidneys, which expel it from the body.

What is blood made of?

Blood consists of a fluid called plasma in which float billions of red and white blood cells, plus platelets – the cell fragments involved in blood clotting. Blood also contains wastes, nutrients, cholesterol, antibodies, and protein clotting factors that travel within the plasma. The body carefully controls blood temperature, acidity, and salt levels – if these vary too much, blood and body cells could not function properly.

Fluid of life
Besides blood cells, blood is made mainly of plasma – a straw-coloured fluid containing water plus dissolved salts, hormones, fats, sugars, and proteins, as well as tissue wastes.

45% red blood cells
1% white blood cells and platelets
54% plasma

5 MILLION
THE NUMBER OF **RED BLOOD CELLS IN A** DROP **OF BLOOD**

Oxygen transport

Most oxygen is carried within the red blood cells. A small amount of oxygen also dissolves in plasma. After a red blood cell collects oxygen from the lungs, it takes around 1 minute to complete one circuit around the body. During this circuit, oxygen diffuses into the tissues and carbon dioxide is absorbed into the blood. Oxygen-depleted blood cells are then taken back to the lungs, where the blood releases carbon dioxide and the cycle starts again.

WHERE IS BLOOD MADE?

Strangely, blood is actually manufactured in bone marrow in your flat bones (such as the ribs, sternum, and shoulder blades) – millions of blood cells are produced every single second!

Double circulation
Oxygen-depleted blood is pumped from the right side of the heart to the lungs. Blood rich in oxygen from the lungs is pumped from the left side of the heart out to the body.

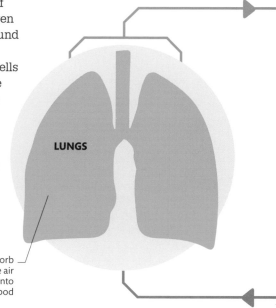

LUNGS

Lungs absorb oxygen from the air and releases it into the blood

What the body needs

All the living cells throughout the body needs various things to help them function properly. Blood carries these vital supplies, such as oxygen, salts, fuel (in the form of glucose or fats), and protein building blocks – amino acids – for growth and repair. Blood also carries hormones, such as adrenalin, which are chemicals that affect the behaviour of cells.

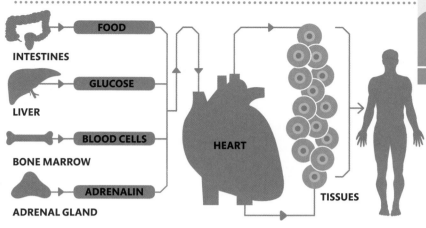

INTESTINES — FOOD

LIVER — GLUCOSE

BONE MARROW — BLOOD CELLS

ADRENAL GLAND — ADRENALIN

HEART

TISSUES

What the body doesn't need

Wastes, such as lactic acid, are produced as by-products of normal cell function. Blood quickly carries the wastes away to prevent imbalances. Some wastes may be transported to the kidneys, to be expelled in urine, or can be carried to the liver to be converted back into something that the cells need.

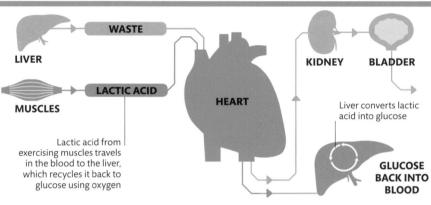

LIVER — WASTE

MUSCLES — LACTIC ACID

Lactic acid from exercising muscles travels in the blood to the liver, which recycles it back to glucose using oxygen

HEART

KIDNEY   BLADDER

Liver converts lactic acid into glucose

GLUCOSE BACK INTO BLOOD

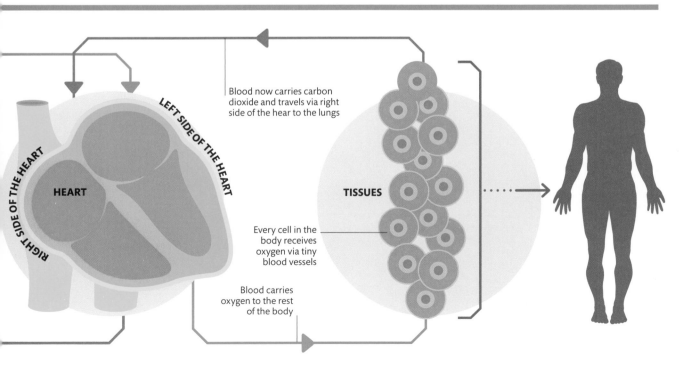

LEFT SIDE OF THE HEART

RIGHT SIDE OF THE HEART

HEART

Blood now carries carbon dioxide and travels via right side of the hear to the lungs

TISSUES

Every cell in the body receives oxygen via tiny blood vessels

Blood carries oxygen to the rest of the body

How the heart beats

The heart is a fist-sized muscular organ that contracts and relaxes around 70 times a minute. This keeps blood flowing around the lungs and body, transporting life-giving oxygen and nutrients.

Heart cycle

Your heart is a muscular pump that is divided into two halves, left and right. Each half of the heart is further divided into two chambers – an upper atrium and a lower ventricle. Valves prevent backflow so that blood keeps travelling in the correct direction. A patch of heart muscle acts as a natural pacemaker, generating the electrical signal that makes the muscle cycle between contraction and relaxation. The rhythmic squeezing of the heart pumps blood from its right side to the lungs and from the left side to the rest of the body.

ECG recording

Electrical impulses within the heart can be recorded by electrodes to produce an electrocardiogram (ECG). Each heartbeat produces a characteristic trace on the ECG display. Its shape is made up of five phases – P, Q, R, S, and T, each of which is a sign of a particular stage of the heartbeat cycle.

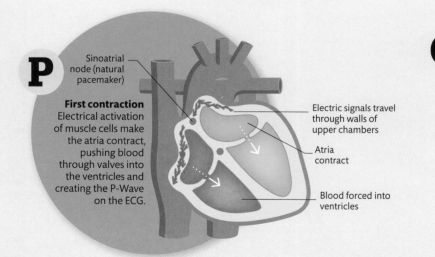

R

Second contraction
The electrical message reaches the tip of the ventricles and spreads throughout the ventricles. The large R-Wave occurs as the powerful, ventricles reach peak contraction.

Ventricles contract

Q

Signal transfer
The electrical signal then passes down the thick, muscular wall between the left and right side of the ventricles, creating the valley of the Q-Wave.

Electricity travels along wall between chambers

P

First contraction
Electrical activation of muscle cells make the atria contract, pushing blood through valves into the ventricles and creating the P-Wave on the ECG.

Sinoatrial node (natural pacemaker)

Electric signals travel through walls of upper chambers

Atria contract

Blood forced into ventricles

WHAT CREATES THE SOUND OF THE HEARTBEAT?

The heart has four valves, and the opening and closing, in pairs, of these heart valves produces the familiar lub-dub sound of the heartbeat.

How electrical signals travel

The heart's pacemaker, the sinoatrial node, is a region of muscle in the upper right atrium. It starts a regular electrical impulse that is conducted throughout the heart by specialized nerve fibres. Heart muscle cells are adept at spreading electrical messages rapidly, so the heart muscle contracts in an orderly sequence, first the two atria followed by the two ventricles.

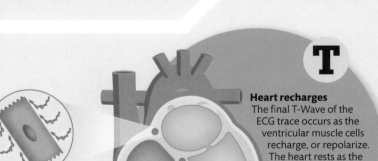

Oxygen rich blood from the lungs are pumped to the rest of the body

Atrium relaxed

S

Electricity travels back

The S-Wave and flat ST segment occur as the ventricles are contracting and emptying of blood. The atrial muscle cells have recharged, ready for the next contraction.

Electricity travels back up towards atria

Blood from heart's right side is pumped to the lungs

Ventricles still contracted

Specialized cells
Natural pacemaker cells in the heart are "leaky" and allow a flow of ions (charged particles) in and out. This generates a regular electrical impulse that causes the heart to beat. Heart (cardiac) muscle cells have branched fibres that let electrical messages spread quickly to the neighbouring muscle cells.

Natural pacemaker

Electrical current

Cardiac muscle cell

T

T

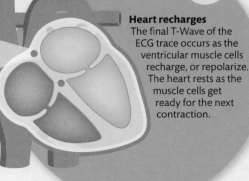

Heart recharges
The final T-Wave of the ECG trace occurs as the ventricular muscle cells recharge, or repolarize. The heart rests as the muscle cells get ready for the next contraction.

HEART MUSCLE CELLS RECHARGE

S

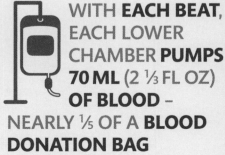

WITH **EACH BEAT**, EACH LOWER CHAMBER **PUMPS 70 ML (2 ⅓ FL OZ) OF BLOOD** – NEARLY ⅕ OF A **BLOOD DONATION BAG**

How blood travels

Blood travels through arteries, capillaries, and veins. Arteries have muscular, elastic walls to even out surges in pressure as the heart pumps. Veins have thinner walls and can distend to help lower blood pressure. If blood pressure rises too high, damage increases the risk of a heart attack or a stroke.

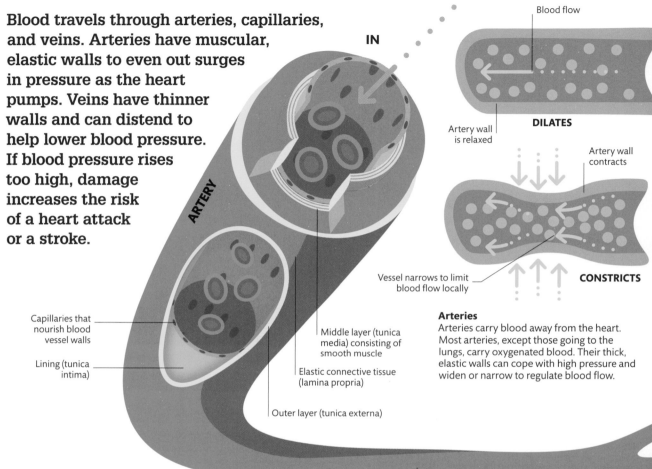

IN

ARTERY

Blood flow

DILATES

Artery wall is relaxed

Artery wall contracts

Vessel narrows to limit blood flow locally

CONSTRICTS

Capillaries that nourish blood vessel walls

Lining (tunica intima)

Middle layer (tunica media) consisting of smooth muscle

Elastic connective tissue (lamina propria)

Outer layer (tunica externa)

Arteries
Arteries carry blood away from the heart. Most arteries, except those going to the lungs, carry oxygenated blood. Their thick, elastic walls can cope with high pressure and widen or narrow to regulate blood flow.

Artery splits into narrower arterioles

Blood pressure
The arteries pulse with blood in time with the heartbeat and so the pressure inside them rises and falls in waves. Arterial pressure is greatest just after the heart contracts (systolic blood pressure) and is lowest when the heart rests between beats (diastolic blood pressure). Pressure is much lower in the capillaries as they are so numerous they spread the force widely. Once blood reaches the veins, its pressure is minimal.

Ranges of pressure
Blood pressure is measured in millimetres of mercury (mmHg) and typical blood pressure varies rhythmically between 120 and 80 mmHg. Although the pressure is lower in both capillaries and veins, blood pressure never reaches 0 mmHg.

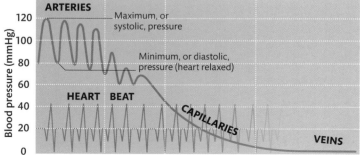

ARTERIES

Blood pressure (mmHg)

120
100
80
60
40
20
0

Maximum, or systolic, pressure

Minimum, or diastolic, pressure (heart relaxed)

HEART BEAT

CAPILLARIES

VEINS

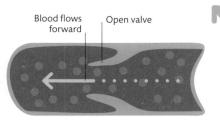

Blood flows forward · Open valve

VALVE OPEN

Closed valve · Blood cannot flow back

VALVE CLOSED

OUT

Route through the body

Blood pulses away from the heart in large arteries, which divide to form smaller arterioles. From the arterioles, blood enters a network of capillaries. In lung capillaries, blood collects oxygen and releases carbon dioxide gas. In body capillaries, blood releases oxygen and collects carbon dioxide. Blood then flows into venules, which join up to form veins returning blood to the heart.

VEIN

Layer of smooth muscle

Lamina propria

Valve

Tunica intima

Veins

Veins carry blood back to the heart. Pressure in them is very low (5–8 mmHg) and the long veins in the legs have a one-way valve system to stop back flow due to gravity.

CAPILLARIES

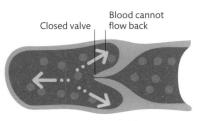

Capillaries

Capillaries form an extensive network that branches finely through body tissues. The entrance to some capillaries is protected by muscle rings (sphincters), which can shut down that part of the network.

Small venules join up to form a larger vein

Small venules

Measuring blood pressure

To measure your blood pressure, a medic inflates a cuff around your arm until the pressure is high enough to stop arterial blood flow. Pressure is then slowly released until blood can just squirt past the cuff, producing a distinct sound that pinpoints systolic blood pressure. As cuff pressure continues to fall, sounds suddenly stop at the point where blood flow is no longer constricted, which pinpoints diastolic blood pressure.

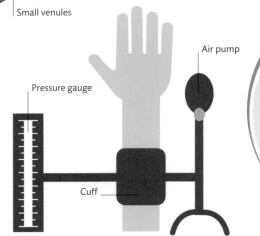

Pressure gauge

Air pump

Cuff

WHY IS HIGH BLOOD PRESSURE SO HARMFUL?

High blood pressure damages artery linings. This can trigger a build-up of cholesterol-laden plaque, which hastens hardening and furring up of the arteries.

Broken blood vessels

Blood vessels permeate the tissues of the body. Their thin walls allow oxygen and nutrients to pass but are easily damaged. Repair systems allow blood to clot so that any damage is quickly fixed, but sometimes unwanted clotting causes a blockage.

Bruising

When a part of the body is knocked, tiny blood vessels may rupture and leak blood into surrounding tissues. Some people bruise more easily than others, especially the elderly. This is sometimes related to blood clotting disorders or nutrient deficiency such as lack of vitamin K (needed to make clotting factors) or vitamin C (needed to make the protein collagen).

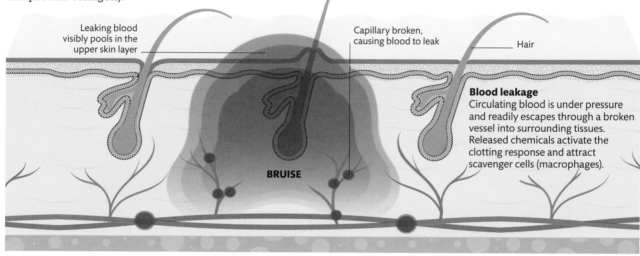

Leaking blood visibly pools in the upper skin layer

Capillary broken, causing blood to leak

Hair

BRUISE

Blood leakage
Circulating blood is under pressure and readily escapes through a broken vessel into surrounding tissues. Released chemicals activate the clotting response and attract scavenger cells (macrophages).

Clotting

A damaged blood vessel must be sealed quickly to prevent blood loss. A complex sequence of reactions causes inactive proteins dissolved within the blood to activate and plug the damage. The blood vessel may constrict to slow blood flow and reduce blood loss from the circulation.

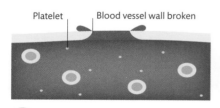

Platelet   Blood vessel wall broken

1 Initial opening
Exposure of proteins such as collagen in a broken blood vessel wall immediately attracts cell fragments called platelets.

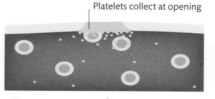

Platelets collect at opening

2 Forming a clot
Platelets stick together and release chemicals that trigger fibrin – a protein circulating in the blood – to form fibres.

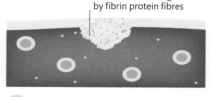

Platelets bound together by fibrin protein fibres

3 Holding the clot
A sticky web of fibrin fibres forms a net that binds platelets together. The web traps red blood cells to form a clot.

How bruises heal

Bruises start purple – the colour of oxygen-poor blood cells seen under the skin. Scavenging macrophage cells recycle the spilled red blood cells as they clean up the area, converting the red blood pigments into first green, then yellow pigments.

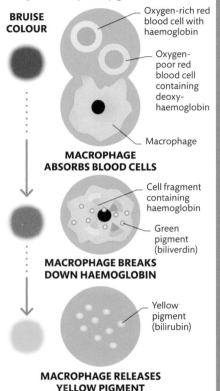

BRUISE COLOUR

Oxygen-rich red blood cell with haemoglobin

Oxygen-poor red blood cell containing deoxy-haemoglobin

Macrophage

MACROPHAGE ABSORBS BLOOD CELLS

Cell fragment containing haemoglobin

Green pigment (biliverdin)

MACROPHAGE BREAKS DOWN HAEMOGLOBIN

Yellow pigment (bilirubin)

MACROPHAGE RELEASES YELLOW PIGMENT

Varicose veins

Varicose veins are a price we pay for walking on two legs rather than four. Valves in the long leg veins let blood travel up against gravity. In surface veins, these valves can collapse, and blood pools, forming bulges. Varicose veins may be hereditary and may also result from increased pressure during pregnancy.

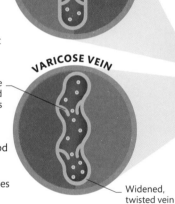

HEALTHY VEIN

Blood restricted from flowing backwards

Healthy vein
A series of valves stops blood from flowing backwards. This allows blood to flow up the length of the leg against the pull of gravity.

VARICOSE VEIN

Valve turned inside out, allowing blood to leak backwards

Pressure builds
When weak valves give way, gravity causes blood to fall backwards and pool in the veins. Increased pressure causes the veins to become dilated and twisted.

Widened, twisted vein

Clot broken up and dispersed by enzymes

Blood vessel wall repaired

4 Clot dissolves
Cells that repair the wound also release enzymes that slowly break down the platelet/fibrin clot – a process called fibrinolysis.

Blocked blood vessels

Raised blood pressure or high glucose levels slowly damage artery walls. Platelets stick to injured areas to fix the damage. If blood cholesterol levels are also high, this seeps into affected areas, causing a build-up that narrows the artery and restricts blood flow. If arteries supplying heart muscle are affected, it can cause a heart attack. When blood flow to the brain is reduced, the memory is affected.

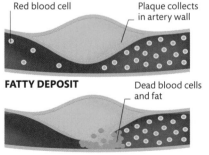

Red blood cell

Plaque collects in artery wall

FATTY DEPOSIT

Dead blood cells and fat

BLOCKED BLOOD VESSEL

Limiting blood flow
Fatty deposits may collect in damaged areas in arteries to form plaques. These deposits cause the arteries to narrow and stiffen, restricting blood flow.

Heart problems

The heart is a vital organ – if it stops pumping blood, cells will not receive the oxygen and nutrients they need. Without oxygen or glucose, brain cells cannot function and you lose consciousness.

Vulnerable vessels

Heart muscle needs more oxygen than any other muscle in the body and the heart has its own coronary arteries to supply its needs as it cannot absorb oxygen from the blood in its chambers. The left and right coronary arteries are relatively narrow and prone to hardening and furring up (narrowing) – a potentially life-threatening process known as atherosclerosis.

IS LAUGHTER REALLY THE BEST MEDICINE?

It may very well be true – laughter can increase your blood flow and relax your blood vessel walls.

Constricted blood flow
A narrowing of a coronary artery may be caused by a build up of fatty deposits (see p.127).

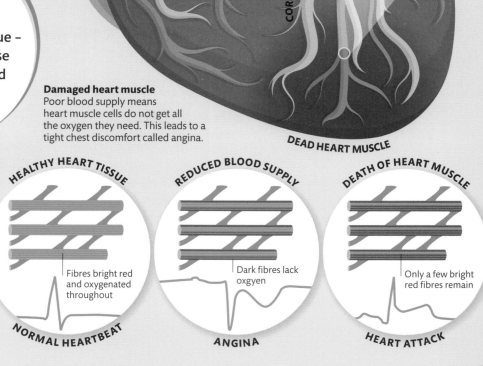

AORTA

CORONARY ARTERY

CORONARY SINUS

Blood cell

Plaque in artery

DEAD HEART MUSCLE

Damaged heart muscle
Poor blood supply means heart muscle cells do not get all the oxygen they need. This leads to a tight chest discomfort called angina.

Decreasing oxygen supply
The heart has specialized cardiac muscle cells whose branched fibres spread electrical messages quickly. Characteristic changes on an ECG (electrocardiogram) help doctors to diagnose whether chest pain is due to poor blood supply (angina) or muscle cell death (heart attack).

HEALTHY HEART TISSUE
Fibres bright red and oxygenated throughout
NORMAL HEARTBEAT

REDUCED BLOOD SUPPLY
Dark fibres lack oxgyen
ANGINA

DEATH OF HEART MUSCLE
Only a few bright red fibres remain
HEART ATTACK

Heart rhythm problems

If the heart is beating too fast, too slowly, or irregularly, medics say that it has an arrhythmia, or abnormal heart rhythms. Most arrhythmias are harmless, such as premature extra beats that feel like a flutter or skipped heart beat. Atrial fibrillation is the most common type of serious arrhythmia, in which the two upper chambers of the heart (atria) beat irregularly and fast. This can cause dizziness, shortness of breath, and fatigue and also increases the risk of suffering a stroke. Some arrhythmias can be treated with drugs. Some need defibrillation to reset and normalize electrical activity.

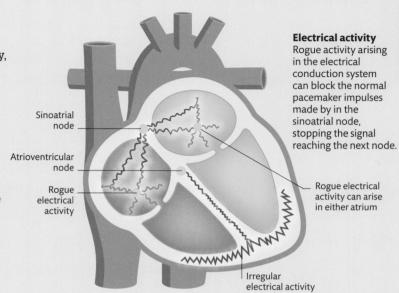

Electrical activity
Rogue activity arising in the electrical conduction system can block the normal pacemaker impulses made by in the sinoatrial node, stopping the signal reaching the next node.

Sinoatrial node

Atrioventricular node

Rogue electrical activity

Rogue electrical activity can arise in either atrium

Irregular electrical activity

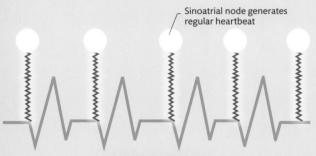

Sinoatrial node generates regular heartbeat

NORMAL HEART BEAT

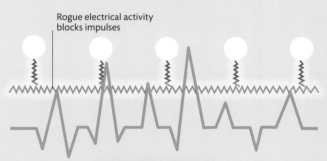

Rogue electrical activity blocks impulses

IRREGULAR HEART BEAT

Electrical interference
The coordinated beating of the heart relies on a clear signal reaching the ventricles from the sinoatrial node. If rogue electrical activity gets in the way, the heart's rhythm of contraction is disturbed and can become erratic.

THE HUMAN HEART BEATS MORE THAN 36 MILLION TIMES A YEAR – ABOUT 2.8 BILLION TIMES IN AN AVERAGE LIFETIME

DEFIBRILLATION

Some life-threatening arrhythmias can be treated by defibrillation. A burst of electricity is delivered to the chest in an attempt to re-establish normal heart electrical activity and contraction. Defibrillation only works if a "shockable" rhythm is present, such as ventricular fibrillation. It cannot restart the heart if no electrical activity is detected (asystole). Cardiopulmonary resuscitation can trigger electrical activity so that defibrillation can be tried.

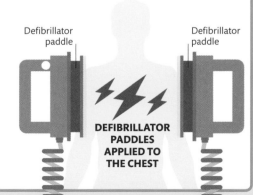

Defibrillator paddle

Defibrillator paddle

DEFIBRILLATOR PADDLES APPLIED TO THE CHEST

Exercising and its limits

When you go for a jog or a sprint, extra blood is pumped to your muscles, providing you with the vital ingredient to make energy – oxygen. Deep, regular breaths replenish your muscles with oxygen and set your pace.

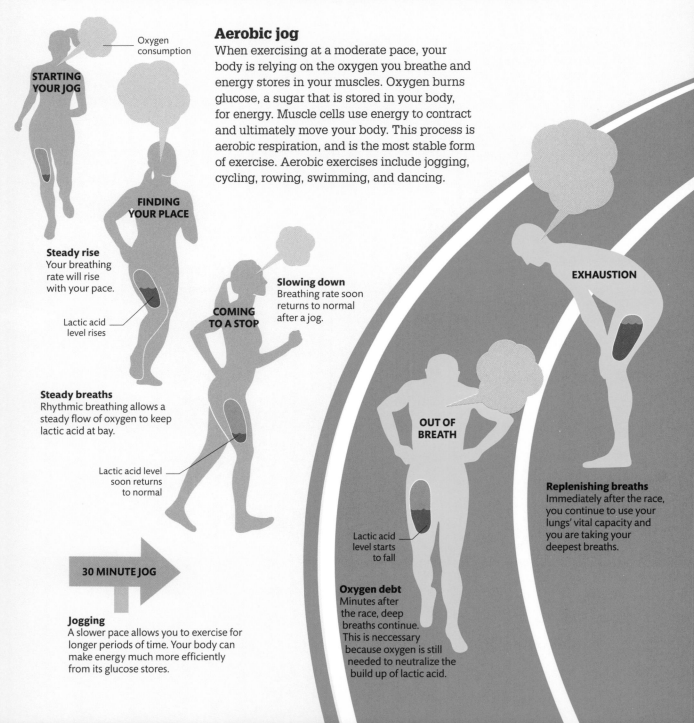

STARTING YOUR JOG

Oxygen consumption

Aerobic jog

When exercising at a moderate pace, your body is relying on the oxygen you breathe and energy stores in your muscles. Oxygen burns glucose, a sugar that is stored in your body, for energy. Muscle cells use energy to contract and ultimately move your body. This process is aerobic respiration, and is the most stable form of exercise. Aerobic exercises include jogging, cycling, rowing, swimming, and dancing.

FINDING YOUR PLACE

Steady rise
Your breathing rate will rise with your pace.

Lactic acid level rises

COMING TO A STOP

Slowing down
Breathing rate soon returns to normal after a jog.

EXHAUSTION

Steady breaths
Rhythmic breathing allows a steady flow of oxygen to keep lactic acid at bay.

Lactic acid level soon returns to normal

OUT OF BREATH

Lactic acid level starts to fall

Replenishing breaths
Immediately after the race, you continue to use your lungs' vital capacity and you are taking your deepest breaths.

30 MINUTE JOG

Jogging
A slower pace allows you to exercise for longer periods of time. Your body can make energy much more efficiently from its glucose stores.

Oxygen debt
Minutes after the race, deep breaths continue. This is neccessary because oxygen is still needed to neutralize the build up of lactic acid.

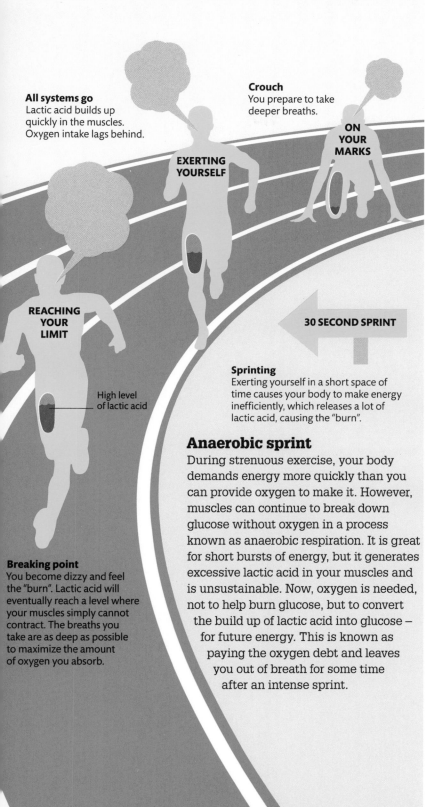

All systems go
Lactic acid builds up quickly in the muscles. Oxygen intake lags behind.

EXERTING YOURSELF

Crouch
You prepare to take deeper breaths.

ON YOUR MARKS

REACHING YOUR LIMIT

High level of lactic acid

30 SECOND SPRINT

Sprinting
Exerting yourself in a short space of time causes your body to make energy inefficiently, which releases a lot of lactic acid, causing the "burn".

Anaerobic sprint

During strenuous exercise, your body demands energy more quickly than you can provide oxygen to make it. However, muscles can continue to break down glucose without oxygen in a process known as anaerobic respiration. It is great for short bursts of energy, but it generates excessive lactic acid in your muscles and is unsustainable. Now, oxygen is needed, not to help burn glucose, but to convert the build up of lactic acid into glucose – for future energy. This is known as paying the oxygen debt and leaves you out of breath for some time after an intense sprint.

Breaking point
You become dizzy and feel the "burn". Lactic acid will eventually reach a level where your muscles simply cannot contract. The breaths you take are as deep as possible to maximize the amount of oxygen you absorb.

Reaching your limit

A build up of lactic acid in your body is the reason why you get tired during exercise. Lactic acid interferes with muscle contraction, which results in physical exhaustion. Oxygen is needed to get rid of lactic acid, which is why you breathe heavily after exercise. This build up of lactic acid happens during both aerobic and anaerobic exercise, but it occurs quicker in the latter. Brain cells can only burn glucose for fuel and as exercising muscles deplete the body's available glucose supplies, mental fatigue also sets in.

EFFECT OF LACTIC ACID IN MUSCLES

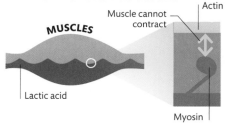

MUSCLES

Muscle cannot contract

Actin

Lactic acid

Myosin

HYDRATION

Drinking water during exercise helps regulate body temperature through sweating and flushes away lactic acid. Water in blood plasma is sweated out, so your blood thickens and your heart works harder to pump blood around the body. This is called cardiac drift, and it's one reason why you can't respire aerobically and jog forever.

FULLY HYDRATED: 75%

LIMIT OF SAFE DEHYDRATION: 70%

Fitter and stronger

Exercise that makes your heart race and your lungs breathe hard and deep is called cardiovascular – it strengthens the heart and improves stamina. In contrast, exercise that forces you to contract muscles repetitively is called resistance training, and it can build and strengthen your muscles.

Cardiovascular exercise

When you perform cardiovascular exercise, such as jogging, swimming, cycling, or brisk walking, you train your cardiovascular system. Your heart rate climbs, beating faster in order to pump more blood around your body, especially to the chest muscles that influence the depth of your breaths. As your body's demand for oxygen rises, your breathing rate and depth rises accordingly. Your blood is saturated with as much oxygen as possible to provide your body with the energy it needs.

Chest muscles

Muscles within the neck, chest wall, abdomen, and back co-ordinate to expand and reduce the size of your ribcage, so that the volume of air your lungs inhale and exhale increases.

Deep breaths include red and blue areas

Lung capacity

Your tidal volume is the volume of air that flows into your lungs during a relaxed breath. If you try to breathe all the air out of your lungs, some air remains as your residual volume, and cannot be breathed out. Your vital capacity, the deepest breath you can take when training, is the rest of your lung volume excluding the residual volume.

VITAL CAPACITY

TIDAL VOLUME

RESIDUAL VOLUME

Air that remains in lungs after deep breath

Relaxed breathing

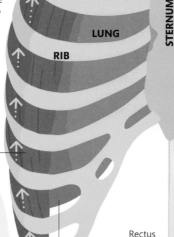

Scalene muscles contract to raise upper ribs

Internal intercostal muscles contract and tilt ribs downwards

Lung volume reduced as muscles contract and ribs tilt

COLLARBONE

STERNUM

LUNG

RIB

External intercostal muscles contract and tilt ribs upwards

Rectus abdominis muscle pulls ribcage downwards

Lung volume increases due to ribs tilting upwards

External oblique muscles contract and shorten to pull ribs downwards

BREATHING IN

BREATHING OUT

WHICH TYPE OF EXERCISE BURNS MORE FAT?

It depends on the individual, but a combination of both cardio and weight training will result in greater fat loss than just doing one or the other.

Resistance training

Weight training builds your muscle, but so does dancing, gymnastics, and yoga – they are all forms of resistance training. A repetition (rep) is one complete motion of exercise. A set is a group of consecutive reps that will contract a particular muscle, or multiple muscles, repeatedly. You can target muscles to grow by choosing to perform a selection of sets and reps over a period of time. The fewer reps you are able to do per set, the tougher your workout.

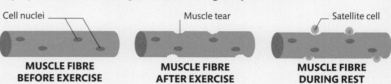

Cell nuclei

MUSCLE FIBRE BEFORE EXERCISE

Muscle tear

MUSCLE FIBRE AFTER EXERCISE

Satellite cell

MUSCLE FIBRE DURING REST

DOING A REP

Rectus abdominis muscle

Bow pose
Yoga is a good way to grow muscle steadily. The bow pose forces the rectus abdominis muscle to contract and tear slightly. Repeating this as a "rep" will start the muscle growth process.

Muscle growth process
Exercise tears muscle fibres, which are then repaired by satellite cells. Although muscle fibres are single body cells, they have many nuclei, and they incorporate the satellite cells, along with their nuclei – growing as they do so. During a break from exercise, your muscle fibres shrink, but they retain the nuclei from the satellite cells and regain their size quickly after re-training.

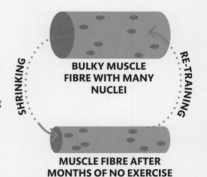

SHRINKING

RE-TRAINING

BULKY MUSCLE FIBRE WITH MANY NUCLEI

MUSCLE FIBRE AFTER MONTHS OF NO EXERCISE

RATES OF EXERCISE

Exercise intensity can be expressed as the percentage of your maximum heart rate. When you go for a jog, you are working your heart at about 50 per cent of its potential power. Athletes who have reached their peak fitness can work their heart at maximum strength – 100 per cent. A fitness instructor can give you a target heart rate to reach when training (which varies with age) while achieving your fitness goals.

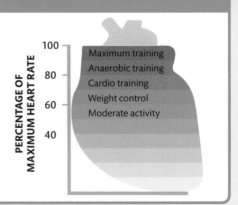

PERCENTAGE OF MAXIMUM HEART RATE

100
80
60
40

Maximum training
Anaerobic training
Cardio training
Weight control
Moderate activity

WHEN YOU **SLEEP, HORMONES** THAT **STIMULATE MUSCLE GROWTH** ARE RELEASED

Maximizing your fitness

While exercising is necessary to maintain health, regular training can improve your overall fitness. Your body will adapt to tough training regimes; muscles get thicker, breaths get deeper, and your state of mind is even enhanced.

Positive results of regular exercise

If you exercise regularly, you will see widespread improvements across your body. Adults benefit from just 30 minutes of brisk exercise on most days, while children need at least 60 minutes of running around. Keeping yourself active is vital for improving your organs and muscles, and by exerting yourself in steady bouts your body systems will become more efficient and eventually will start to function at the best of their ability.

BRAIN

HEART

LUNG

LIVER

OXYGEN INTAKE

Depth of each breath increases with exercise

Exercise strengthens your chest muscles, which allows greater lung expansion. So, the amount of air your lungs can hold increases, and your breathing rate rises, resulting in a greater amount of oxygen absorbed when exercising and also at rest.

ARTERY DIAMETER INCREASE

Artery widens

When exercising, nerve signals cause arteries to dilate, or widen, increasing blood flow. This delivers more oxygenated blood to the muscles. If you exercise regularly, the diameter that your arteries dilate to when you exercise becomes wider, maximizing the amount of oxygen that reaches your muscles.

METABOLIC SYSTEMS IMPROVE

Metabolic process occuring in liver

Your metabolic rate is the speed at which chemical processes, such as digestion or the burning of fat, take place in your body. Exercise generates heat, which speeds up these processes in your organs, even after you finish exercising.

COGNITIVE IMPROVEMENT

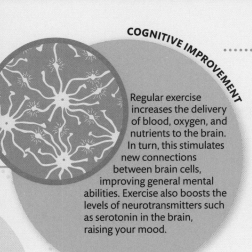

Regular exercise increases the delivery of blood, oxygen, and nutrients to the brain. In turn, this stimulates new connections between brain cells, improving general mental abilities. Exercise also boosts the levels of neurotransmitters such as serotonin in the brain, raising your mood.

STRONGER CARDIAC MUSCLE

Cardiac muscle fibres grow in size, but not via satellite cells as is the case in muscles in the rest of your body. Instead, their existing fibres grow stronger. Your heart's contractions become stronger too, and it distributes blood more thoroughly around the body, lowering your resting heart rate.

STRONGER MUSCLES

Having strong muscles increases your physical strength, strengthens your bones, improves posture, flexibility, and how much energy you burn during exercise and while at rest. Strong muscle is also more resilient to exercise-induced injury.

Reaching your maximum

During a training program, for most people, the effort you put in reaps great benefits at first, as your fitness increases from your untrained level. Further improvements become ever harder to achieve as you approach your own physiological limits, which depend on age, gender and other genetic factors. You reach your maximum more quickly with a higher intensity training program. The best athletes explore their limits, looking for opportunities to extend them.

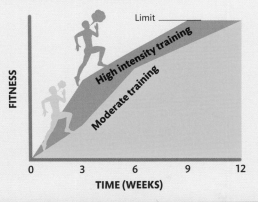

FITNESS

Limit

High intensity training

Moderate training

TIME (WEEKS)
0   3   6   9   12

RESTING HEART RATES

Athletes have low heart rates at rest because training enhances the strength of their cardiac muscle. Compared to those who are untrained, athletes' heart contractions are stronger, and blood is distributed more efficiently with every heartbeat. A trained athlete may have a pulse rate as low as 30–40 beats per minute at rest.

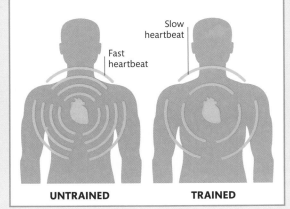

Fast heartbeat

Slow heartbeat

UNTRAINED   **TRAINED**

IN AND
OUT

Feeding the body

Although the body can manufacture many vital chemicals, a lot of the materials we need must be acquired by eating. The energy needed to fuel the body is gained entirely through the food we consume. Once nutrients are absorbed into the bloodstream, they are then transported to different parts of the body where they are put to innumerable tasks.

WHAT IF I DON'T GET WHAT I NEED?

Your body systems will start to fail and you may be afflicted with deficiency diseases. For example, if you do not have enough minerals in your diet, your bones will not grow properly.

Carbohydrates
Carbohydrates are the main energy source for the brain. Whole grains and fruits and vegetables that are high in fibre are healthy sources of carbohydrates.

Water
Some 65 per cent of the body is made up of water. This is constantly being lost through breathing and sweating, and it is critical that it is replenished.

Protein
Proteins are the major structural components of all cells. Healthy protein sources include beans, lean meat, dairy, and eggs.

Sugars

Amino acids

What the body needs

There are six essential types of nutrient that the body needs to get from the diet in order to function properly: fats, protein, carbohydrates, vitamins, minerals, and water. The last three are small enough to be absorbed directly through the lining of the gut, but fats, protein, and carbohydrates need to be broken down chemically into smaller particles before being absorbed. These particles are sugars, amino acids, and fatty acids respectively.

DIGESTIVE TRACT

Fats
Fats are a rich source of energy and help in the absorption of fat-soluble vitamins. Healthier fat sources include dairy, nuts, fish, and vegetable-based oils.

Fatty acids

Vitamins
Vitamins are needed to make things in the body. Vitamin C, for example, is needed to build collagen, which is used in various tissues.

Minerals
Minerals are vital for building bones, hair, skin, and blood cells. They also enhance nerve function and help to turn food into energy.

Building an eye

Every tissue in our body is built and maintained by the nutrients we absorb from our food. The tissues of the human eye, for example, are built from amino acids and fatty acids, and fuelled by sugars. The membranes and spaces are filled with fluids, and vitamins and minerals are needed to convert light into an electrical impulse – the basis of vision itself.

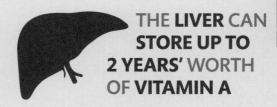

THE **LIVER** CAN STORE UP TO **2 YEARS'** WORTH OF **VITAMIN A**

Cell membranes
All the cells of the eye (and the rest of the body) are surrounded by membranes that are built using fatty acids and proteins.

Energy
The eyes are an extension of the brain, and just like the brain, they need the sugars we get from carbohydrates for energy.

The food of sight
Like all organs of the body, the eye utilizes all six of the essential nutrients. These give it structure and enable it to send visual information to the brain.

Fluids
The eye is filled with fluid, which maintains the pressure in the eye and provides nutrients and moisture to the inner eye tissues. This fluid is 98 per cent water.

Tissue structures
Eyelashes are made up of the protein keratin, which is built from amino acids. Other tissues of the eye are made of the protein collagen.

Vision
Vitamin A is bound to proteins in the eye known as visual pigments. When light hits the cells, the vitamin A changes shape, sending an electrical impulse to the brain.

Red blood cells
The tissues of the eye are oxygenated by the red blood cells, which need the protein haemoglobin and the mineral iron in order to carry oxygen.

How does eating work?

Eating is the process of breaking down food into molecules that are small enough to be absorbed into the bloodstream. For the food, this involves a 9m (30ft) journey through a series of organs known collectively as the gut, or the intestinal tract.

The journey of food

Food begins as a (usually) appetizing meal, and ends up with us taking trips to the toilet. Between these stages, the food has done its job – released its nutrients in a four-stage process involving the mouth, the stomach, the small intestine, and the large intestine. The liver and pancreas also play their roles, as do the hormones leptin and ghrelin. On average, it takes 48 hours for food to pass through the body.

Nutrient absorption
Some nutrients take longer to absorb than others, but most are absorbed in the small intestine.

↑ Vitamins
↑ Sugars
↑ Amino acids
↑ Minerals
↑ Fatty acids
↑ Water
↑ Blood flow

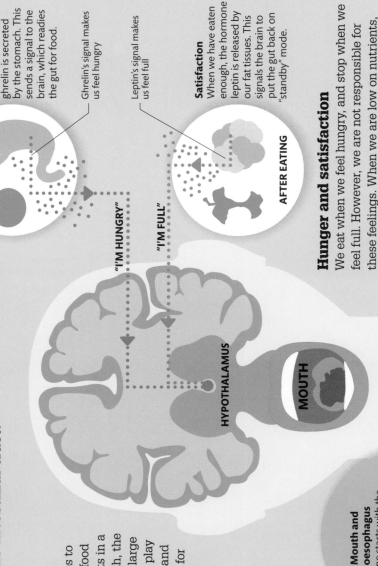

Hunger

A few hours after eating, the hormone ghrelin is secreted by the stomach. This sends a signal to the brain, which readies the gut for food.

Ghrelin's signal makes us feel hungry

BEFORE EATING

Leptin's signal makes us feel full

Satisfaction

When we have eaten enough, the hormone leptin is released by our fat tissues. This signals the brain to put the gut back on "standby" mode.

AFTER EATING

"I'M HUNGRY"

"I'M FULL"

HYPOTHALAMUS

MOUTH

Hunger and satisfaction

We eat when we feel hungry, and stop when we feel full. However, we are not responsible for these feelings. When we are low on nutrients, the hormone ghrelin is released by the stomach, making us feel hungry – and when we are full, the hormone leptin is released by our fat tissues, inhibiting our appetites.

BLOODSTREAM

OESOPHAGUS

1 Mouth and oesophagus

Stage one starts with the mechanical breakdown of food by chewing. This mixes the food with saliva, which begins to digest it chemically. The food is then swallowed, which drops it into the oesophagus (see p.142).

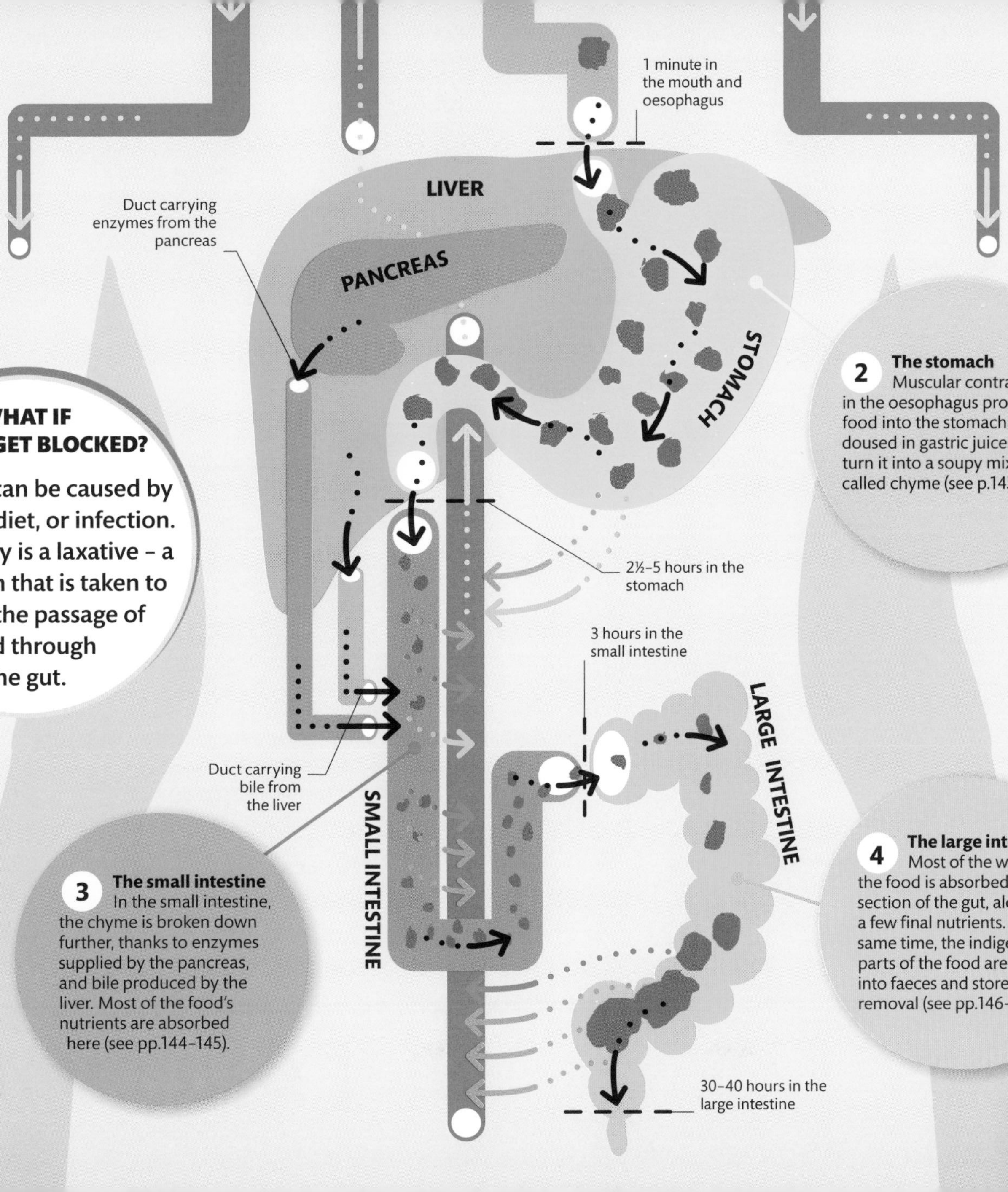

STOMACH

LARGE INTESTINE

LIVER

PANCREAS

SMALL INTESTINE

2 **The stomach**
Muscular contractions in the oesophagus propel the food into the stomach. Here it is doused in gastric juices, which turn it into a soupy mixture called chyme (see p.143).

4 **The large intestine**
Most of the water from the food is absorbed in this last section of the gut, along with a few final nutrients. At the same time, the indigestible parts of the food are pressed into faeces and stored for removal (see pp.146–147).

3 **The small intestine**
In the small intestine, the chyme is broken down further, thanks to enzymes supplied by the pancreas, and bile produced by the liver. Most of the food's nutrients are absorbed here (see pp.144–145).

1 minute in the mouth and oesophagus

2½–5 hours in the stomach

3 hours in the small intestine

30–40 hours in the large intestine

Duct carrying enzymes from the pancreas

Duct carrying bile from the liver

WHAT IF THINGS GET BLOCKED?

Blockages can be caused by stress, bad diet, or infection. One remedy is a laxative – a medication that is taken to smooth the passage of food through the gut.

A mouth to feed

The long and convoluted journey taken by food through the body begins with a brief stay in the mouth and an acid bath in the stomach. The goal of this first stage of digestion is to turn food into chyme – a soup of nutrients that is then moved on to the small intestine for processing.

Heading south

The route from mouth to stomach is a vertical one, via a connecting tube called the oesophagus. The food is propelled by gravity and by muscular contractions in the oesophagus known as peristaltic waves.

Chewing
When food is in the mouth, the epiglottis stands up to keep the windpipe open. This allows us to breathe through our noses while chewing.

Air in

Epiglottis up

Swallowing
When we swallow, the epiglottis folds down, closing off the windpipe. At the same time, the soft palate rises to block off the nasal cavity.

Soft palate up

Epiglottis down

Ready to chew again
When food has entered the oesophagus, the epiglottis and soft palate return to their former position. This enables us to breathe and chew again.

Epiglottis up

How to avoid choking

Since we both eat and breathe through our mouths, it is vital that our windpipes can be closed off when we swallow. Luckily, our bodies have an in-built pair of safety devices – a small flap of cartilage in the throat called the epiglottis, and a piece of flexible tissue in the roof of the mouth called the soft palate.

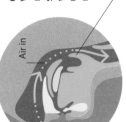

NASAL PASSAGE

SALIVARY GLAND

TONGUE

The salivary glands in the cheeks produce watery saliva

Another salivary gland under the jaw releases saliva at the base of the tongue

OESOPHAGUS

WINDPIPE

Chewing creates a ball of saliva-saturated food

The salivary gland under the tongue produces thick saliva containing enzymes

1 Digestion begins
As food is chewed in the mouth, the salivary glands increase the production of saliva, which helps turn food into a paste. Saliva also contains an enzyme called amylase, which converts starch into more easily absorbable sugars.

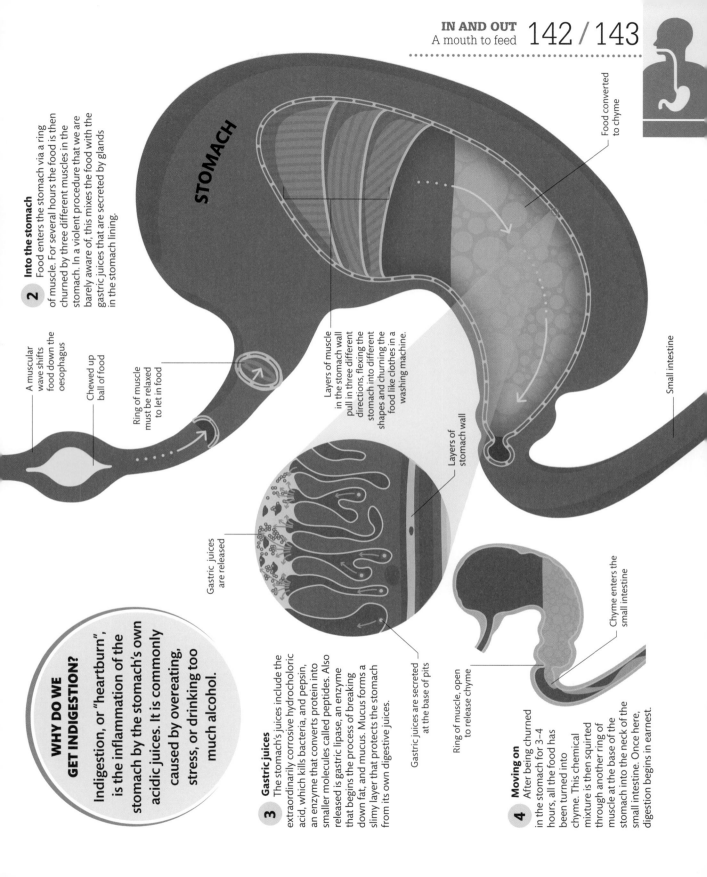

Food converted to chyme

STOMACH

2 **Into the stomach**

Food enters the stomach via a ring of muscle. For several hours the food is then churned by three different muscles in the stomach. In a violent procedure that we are barely aware of, this mixes the food with the gastric juices that are secreted by glands in the stomach lining.

A muscular wave shifts food down the oesophagus

Chewed up ball of food

Ring of muscle must be relaxed to let in food

Layers of muscle in the stomach wall pull in three different directions, flexing the stomach into different shapes and churning the food like clothes in a washing machine.

Layers of stomach wall

Small intestine

Gastric juices are released

Gastric juices are secreted at the base of pits

Ring of muscle, open to release chyme

Chyme enters the small intestine

WHY DO WE GET INDIGESTION?

Indigestion, or "heartburn", is the inflammation of the stomach by the stomach's own acidic juices. It is commonly caused by overeating, stress, or drinking too much alcohol.

3 **Gastric juices**

The stomach's juices include the extraordinarily corrosive hydrochloric acid, which kills bacteria, and pepsin, an enzyme that converts protein into smaller molecules called peptides. Also released is gastric lipase, an enzyme that begins the process of breaking down fat, and mucus. Mucus forms a slimy layer that protects the stomach from its own digestive juices.

4 **Moving on**

After being churned in the stomach for 3–4 hours, all the food has been turned into chyme. This chemical mixture is then squirted through another ring of muscle at the base of the stomach into the small intestine. Once here, digestion begins in earnest.

Gut reaction

Once food has been turned into chyme in the stomach it is squirted into the small intestine. Here, in a frenzy of chemical activity, it is broken down further and finally absorbed by the blood. Each day, around 11.5 litres (20 pints) of food, liquids, and digestive juices pass through the small intestine.

Organs in concert

To help it digest, the small intestine gets help from three other organs: the pancreas, which makes enzymes; the liver, which makes bile; and the bile-storing organ, the gallbladder.

3 Enzyme engine
The pancreas produces three main enzymes: amylase, which turns carbohydrates into sugars; protease, which turns proteins into amino acids; and lipase, which turns fatty droplets into fatty acids.

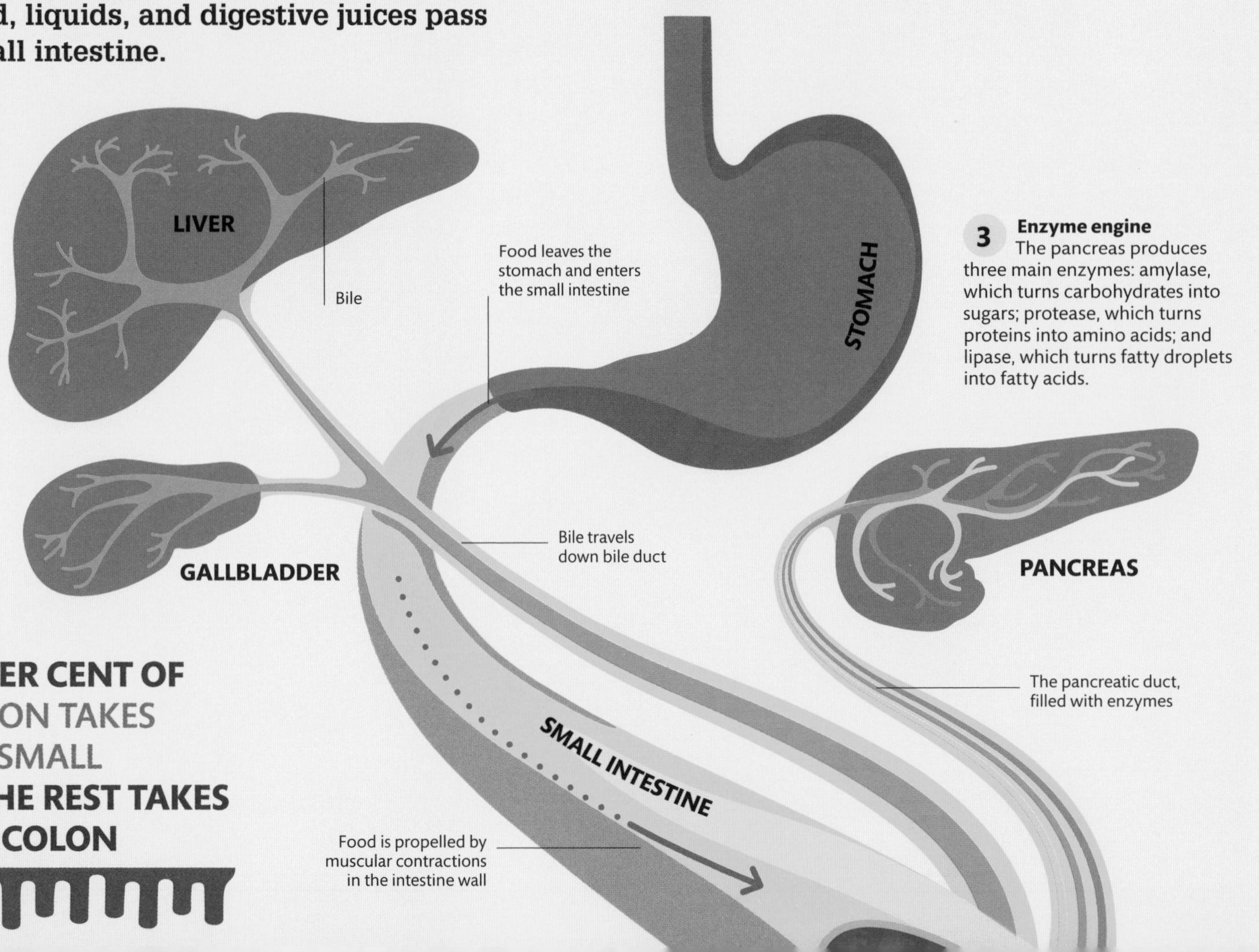

PANCREAS

The pancreatic duct, filled with enzymes

STOMACH

Food leaves the stomach and enters the small intestine

Bile travels down bile duct

Bile

LIVER

SMALL INTESTINE

GALLBLADDER

Food is propelled by muscular contractions in the intestine wall

1 Bile factory
One of the liver's many jobs is to produce bile – a bitter liquid that turns fats into more readily digestible fatty droplets. Once produced, bile is stored in the gallbladder.

2 Bile store
When food leaves the stomach bile leaves the gallbladder and heads for the small intestine. There it mixes with incoming enzymes from the pancreas.

AROUND 95 PER CENT OF ALL ABSORPTION TAKES PLACE IN THE SMALL INTESTINE – THE REST TAKES PLACE IN THE COLON

Opening of ducts carrying digestive juices

Thousands of villi line the intestine wall

4 **Absorption begins**

For 3–5 hours bile and enzymes work together, reducing nutrients to simpler, absorbable forms. Absorption takes place in the intestine wall, which is lined by thousands of finger-like projections. These projections, called villi, greatly increase the surface area of the intestine and so its capacity for absorbing nutrients.

5 **Into the blood**

Villi absorb nutrients and channel them into the blood. Once there, they are taken to the liver and distributed around the body. Meanwhile, the remaining chyme enters the final part of the gut (see pp.146–47). Although not shown here, fat digestion has another complication; on entering the villi, the fatty acids take a trip through the lymph system before finally entering the bloodstream.

BLOODSTREAM

CHEWING THE FAT

Fats are particularly hard to digest. Even after being drenched in hydrochloric acid in the stomach, they are still not fit for enzyme consumption. This is where bile comes in. In a process called emulsion, bile turns fats into fatty droplets, which are then small enough for enzymes to attack.

FAT

FATTY DROPLETS

BILE

ENZYMES

BILE

Amylase digests carbohydrates, producing sugars

CARBOHYDRATE

Sugars

Protease digests protein, producing amino acids

Amino acids

Lipase digests fatty droplets, producing fatty acids

PROTEIN

Fatty acids

FATTY DROPLET

Dissolved sugars

Dissolved amino acids

Dissolved fatty acids

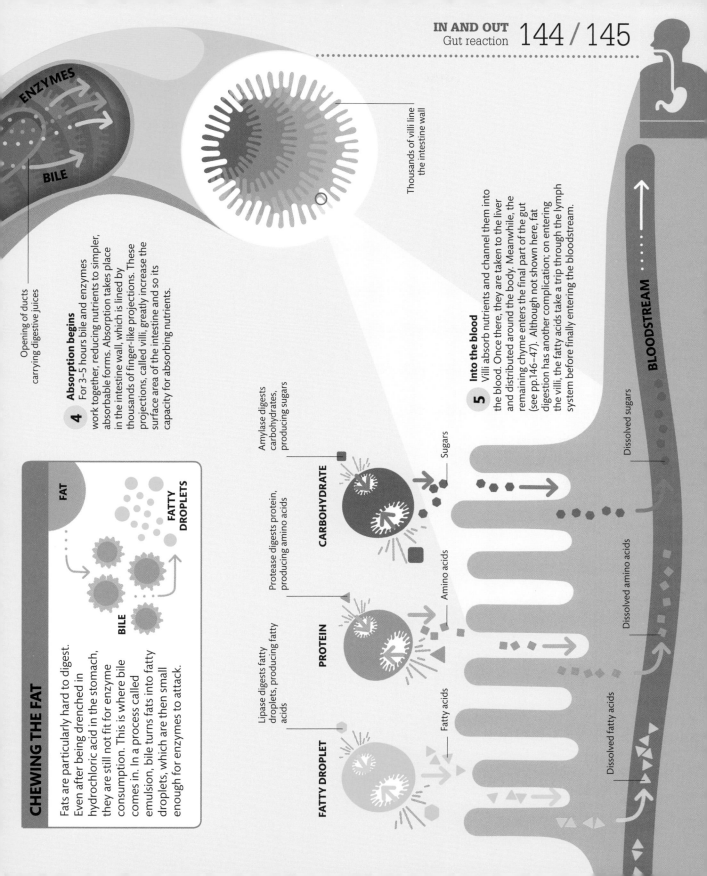

Up, down, and out

The final stage of digestion takes place in the large intestine – a 2.5m- (4ft-) long tube that frames the small intestine. Here, bacteria set to work fermenting carbohydrates, releasing nutrients that are vital for human health. At the same time, faecal matter is compacted, stored, and ejected.

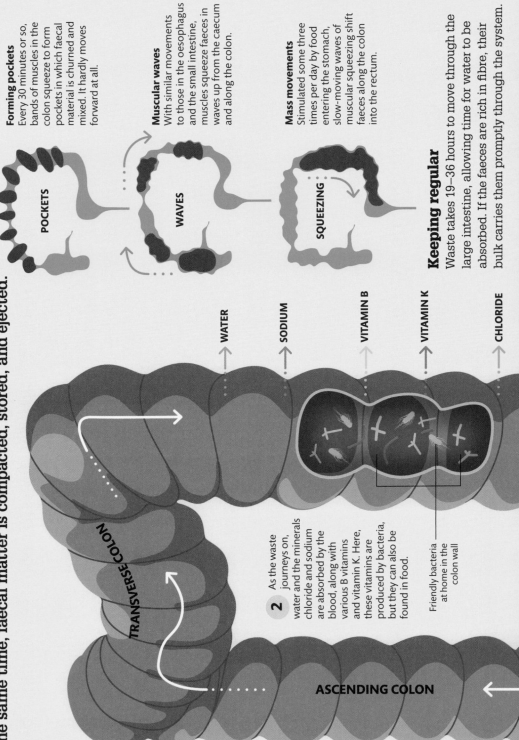

Forming pockets
Every 30 minutes or so, bands of muscles in the colon squeeze to form pockets in which faecal material is churned and mixed. It hardly moves forward at all.

POCKETS

Muscular waves
With similar movements to those in the oesophagus and the small intestine, muscles squeeze faeces in waves up from the caecum and along the colon.

WAVES

Mass movements
Stimulated some three times per day by food entering the stomach, slow-moving waves of muscular squeezing shift faeces along the colon into the rectum.

SQUEEZING

Keeping regular
Waste takes 19–36 hours to move through the large intestine, allowing time for water to be absorbed. If the faeces are rich in fibre, their bulk carries them promptly through the system.

WATER

SODIUM

VITAMIN B

VITAMIN K

CHLORIDE

TRANSVERSE COLON

2 As the waste journeys on, water and the minerals chloride and sodium are absorbed by the blood, along with various B vitamins and vitamin K. Here, these vitamins are produced by bacteria, but they can also be found in food.

Friendly bacteria at home in the colon wall

ASCENDING COLON

WHY DO WE HAVE AN APPENDIX?

The appendix is possibly the remnant of an organ that helped our ancestors digest foliage thousands of years ago. Today, however, it plays no obvious role, except perhaps as a safe refuge for gut bacteria.

WHEN NATURE CALLS

When faeces enter the rectum, stretch receptors trigger a "need to go" reflex by sending impulses to the spinal cord. Motor signals from the spine then tell the internal anal sphincter to relax. At the same time, sensory messages to the brain make a person aware of the need to defecate, and the person makes a conscious decision to relax the external anal sphincter. On a healthy diet, this happens between three times a day and once every three days.

Potassium and bicarbonate are absorbed by the colon to replace the sodium absorbed by the bloodstream

3 Faeces are compacted into the lower colon. They are kept moist by mucus secreted from the colon walls.

DESCENDING COLON

SMALL INTESTINE

Appendix

CAECUM

RECTUM

ANUS

Anus contains both inner and outer sphincters

Journey's end

The large intestine has three main sections: the caecum, where waste from the small intestine is collected; the three-part colon, where nutrients are absorbed; and the rectum, where faeces are expelled. The largest section is the colon, in which colonies of bacteria consume the starches, fibre, and sugars that humans can't digest (see pp.148–49).

1 Having left the small intestine, waste material begins a vertical climb of the caecum.

4 Faeces are expelled via the rectum. Some 60 percent of the volume is made of bacteria, the rest is mostly indigestible fibre.

Bacterial breakdown

Over 100 trillion beneficial bacteria, viruses, and fungi live in the digestive tract. Known collectively as gut microbes, they provide us with nutrients, help us digest, and help us defend against harmful microbes (see pp.172–73).

Swallowing microbes

We receive our first microbes at birth, and more enter our bodies every day of our lives. They enter through the nose and mouth and travel to the stomach, where conditions are too acidic for many to take up permanent residence. The small intestine is likewise too acidic, but many microbes survive just long enough to move into the colon, where they play a vital role in digestion.

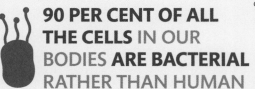

90 PER CENT OF ALL THE CELLS IN OUR BODIES **ARE BACTERIAL** RATHER THAN HUMAN

ANTIBIOTICS

Antibiotics destroy or slow down the growth of bacteria, but they aren't able to discriminate between harmful and friendly bacteria. As a consequence, the friendly microbes in the gut suffer when we take antibiotics. The diversity of gut bacteria starts to decrease as soon as the antibiotic course starts and reaches a minimum about 11 days later. The populations soon bounce back after treatment, but overuse of antibiotics can cause them permanent damage.

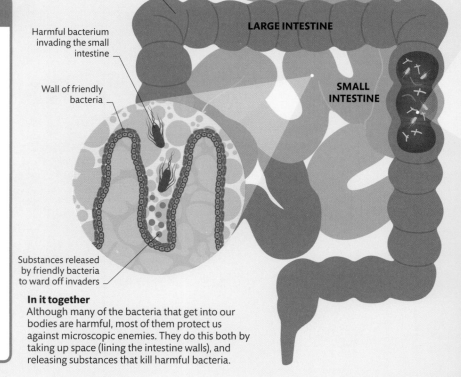

Lactobacilli are common stomach bacteria that are used in probiotic medical treatments. They fight off other bacteria that cause diarrhoea

Helicobacter pylori is a foe, causing ulcers as it burrows into the stomach lining

STOMACH

CHYME

70 per cent of all gut microbes live in the large intestine

LARGE INTESTINE

SMALL INTESTINE

Harmful bacterium invading the small intestine

Wall of friendly bacteria

Substances released by friendly bacteria to ward off invaders

In it together
Although many of the bacteria that get into our bodies are harmful, most of them protect us against microscopic enemies. They do this both by taking up space (lining the intestine walls), and releasing substances that kill harmful bacteria.

Digesting what we can't

The microbes in the colon use the carbohydrates we can't digest for energy. They ferment fibre such as cellulose, which help us absorb dietary minerals such as calcium and iron, are used to produce vitamins, and have other benefits in the body. The microbes themselves also secrete essential vitamins, such as vitamin K.

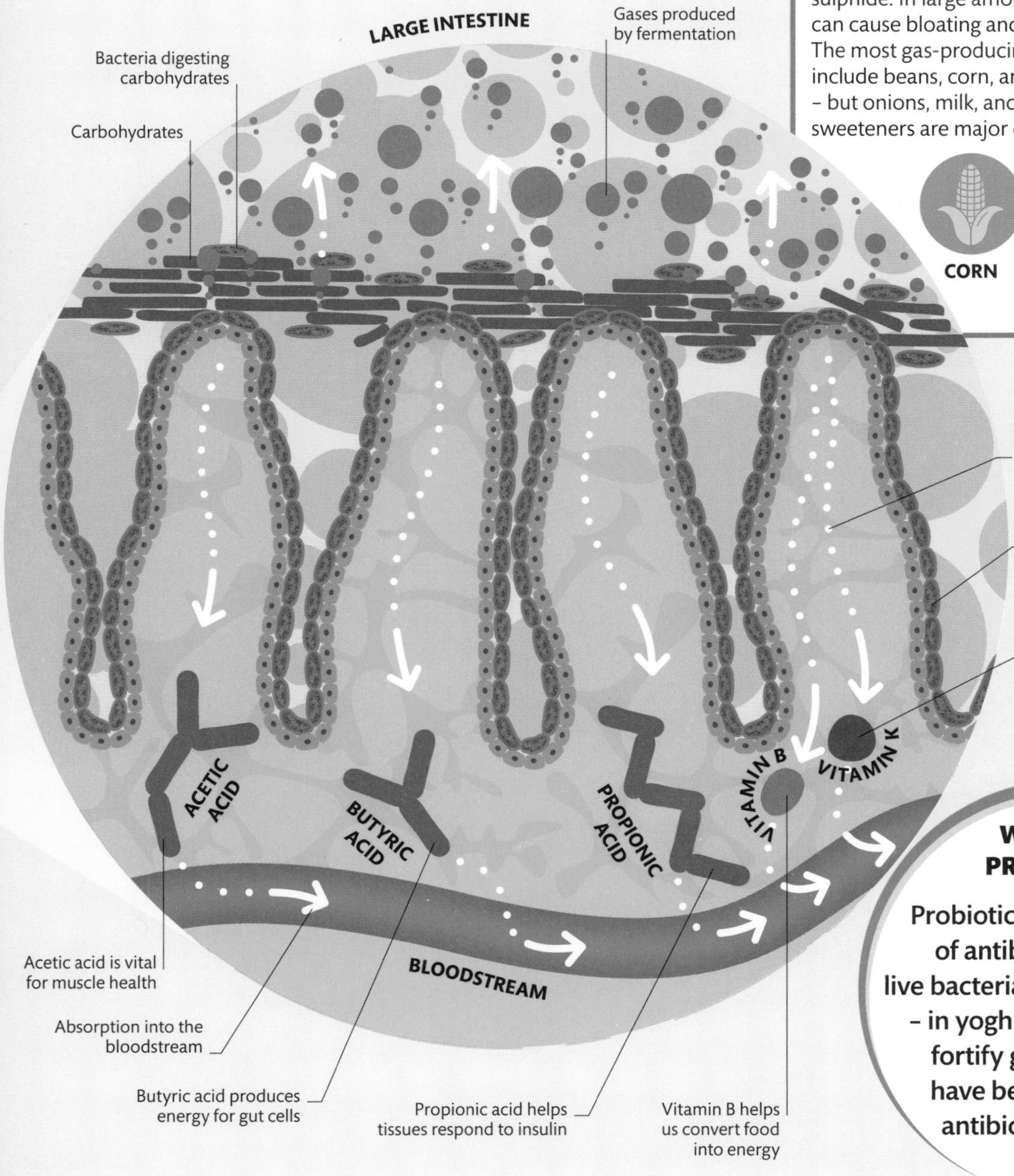

WHAT'S THAT SMELL?

Fermentation by gut microbes produces a number of different gases, including hydrogen, carbon dioxide, methane, and hydrogen sulphide. In large amounts, these can cause bloating and flatulence. The most gas-producing foods include beans, corn, and broccoli – but onions, milk, and artificial sweeteners are major offenders too.

CORN   **BROCCOLI**

LARGE INTESTINE

Gases produced by fermentation

Bacteria digesting carbohydrates

Carbohydrates

Nutrients being absorbed by the large intestine

Wall of friendly bacteria

Vitamin K plays a vital role in blood clotting

ACETIC ACID

BUTYRIC ACID

PROPIONIC ACID

VITAMIN B

VITAMIN K

BLOODSTREAM

Acetic acid is vital for muscle health

Absorption into the bloodstream

Butyric acid produces energy for gut cells

Propionic acid helps tissues respond to insulin

Vitamin B helps us convert food into energy

WHAT ARE PROBIOTICS?

Probiotics are the opposite of antibiotics. They are live bacteria that are consumed – in yoghurts or tablets – to fortify gut bacteria that have been damaged by antibiotics or disease.

Cleaning the blood

As blood travels through the body, it picks up a great deal of waste and excess nutrients. These would quickly reach life-threatening levels without the kidneys, whose job it is to flush them out of the system.

Waterworks

It takes 5 minutes for blood to pass through the kidneys. It enters waste-laden and leaves clean, having passed through countless microscopic filters that turn the waste into urine. The urine then flows to the bladder, at which point we feel the need to urinate. A major component of urine is urea – a waste product formed in the liver (see pp.156–57).

(see pp.156–57)

THE ENTIRE BLOOD STREAM IS FILTERED BY THE KIDNEYS **20–25 TIMES PER DAY**

Each nephron is anchored to the middle part of the kidney, called the medulla

Waste in the form of urine is collected in the medulla

1 Dirty blood in
Waste-laden blood enters the kidney via the renal artery. This artery branches out into a forest of capillaries that feed around a million micro-filters known as nephrons. After being filtered, clean blood leaves the kidney via the renal vein.

Dirty blood flows in

Clean blood flows out

RENAL ARTERY

RENAL VEIN

RENAL PELVIS

MEDULLA

CORTEX

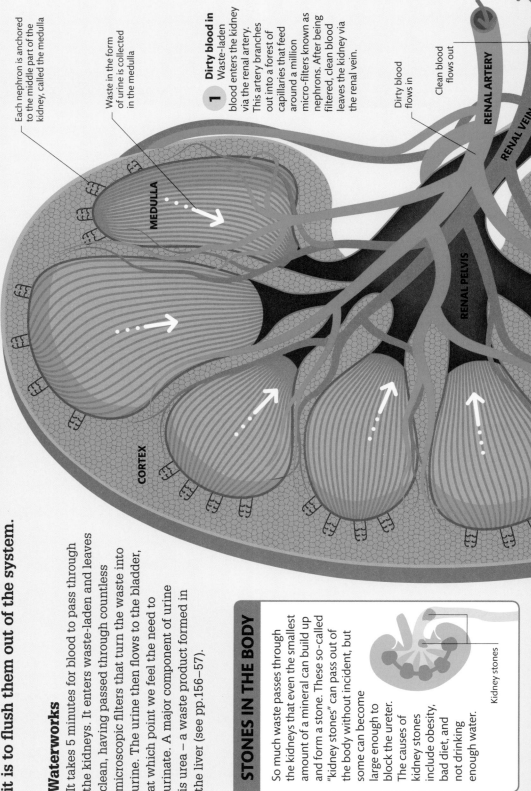

STONES IN THE BODY

So much waste passes through the kidneys that even the smallest amount of a mineral can build up and form a stone. These so-called "kidney stones" can pass out of the body without incident, but some can become large enough to block the ureter. The causes of kidney stones include obesity, bad diet, and not drinking enough water.

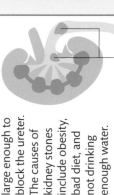

Kidney stones

URETER

MUSCULAR BLADDER WALL

Waste products, including urea, other toxins, and excess salts, flow out in the urine

3 **Collecting urine**
The urine-collecting tubes of the medulla join together as they coverge on the renal pelvis. Here the urine flows past the renal artery and the renal vein and enters a tube called the ureter. The ureter connects the kidney to the bladder.

4 **Waste disposal**
Muscular contractions squeeze the urine along the ureter – which is why our bladders fill even when we are lying down. When the bladder is full, its muscular walls squeeze the urine further, but the urine is halted by a ring of muscle at the base of the bladder. Learning how to control this muscle gives us the choice about when to urinate.

BLADDER

Bladder full of urine

Urethra

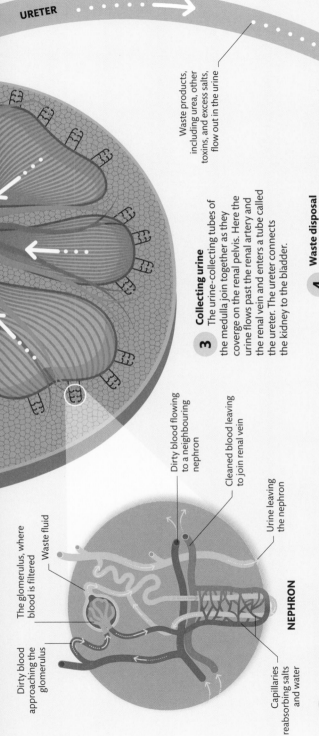

Dirty blood flowing to a neighbouring nephron

Cleaned blood leaving to join renal vein

Urine leaving the nephron

The glomerulus, where blood is filtered

Waste fluid

NEPHRON

Dirty blood approaching the glomerulus

Capillaries reabsorbing salts and water

2 **The filtration process**
As the blood passes through a nephron, it is forced through a tiny filter called a glomerulus, which lets urea and other wastes pass, but keeps blood cells and valuable proteins in the bloodstream. On the far side, the waste fluid passes on a long loop through the kidney, where its composition of salts and water is fine-tuned, before it flows into urine collecting ducts.

WHAT IF THE KIDNEYS FAIL?

If a person's kidneys are too weak to filter their blood, a dialysis machine can be used instead of the kidneys. The person's blood flows through a tube into the machine, gets cleaned and filtered, and then returns to their body.

Water balance

Water levels in the blood have to be kept within a certain range, otherwise the body's cells become too shrunken (dehydrated) or too bloated (overhydrated) to work. For this reason, the kidneys, the endocrine system, and the circulatory system work together to maintain a healthy balance in our bloodstream.

Too little water

We lose water constantly, but there are times when we lose a lot of water quickly – through sweating, vomiting, or diarrhoea, for example. This results in both a decrease in blood volume and a rise in the level of salt relative to water in our blood. These act as triggers for balance to be restored.

1 Low water alert
The hypothalamus receives signals that blood pressure is low and salt levels high. It responds by increasing the production of ADH (antidiuretic hormone), which is carried to the pituitary gland, where it is released into the blood.

Salt detector

Pituitary gland

Stretch receptor on blood vessel warns hypothalamus of decreasing blood pressure

Decreasing water levels in blood vessel

Torrent of ADH

HYPOTHALMUS

LOSING BALANCE

A number of commonly consumed substances uspet our water balance. Alcohol, for example, blocks the pituitary gland from releasing ADH. This means that the kidneys, which are working hard to get rid of the alcohol in the bloodstream, send more water out into the urine. Drinking just one glass of wine can cause the body to get rid of the equivalent of four wine glasses of water. Substances that make us produce a lot of urine are called "diuretics". Caffeine is another diuretic.

Too much water

Far rarer than dehydration is overhydration, which can be caused by extreme water intake after exercise, drug abuse, or disease. This results in an increase in blood volume and a reduction in the level of salt relative to water in the blood.

1 High water alert
The hypothalamus receives signals that blood pressure is high and salt levels low. It responds by producing less ADH. Since ADH instructs the kidneys to store water, a reduction in ADH means an increase in urination.

Stretch receptor on blood vessel warns hypothalamus of increasing blood pressure

Rising water levels in blood vessel

Salt detector

Pituitary gland

HYPOTHALMUS

Trickle of ADH

BRAIN

WATER EXCESS

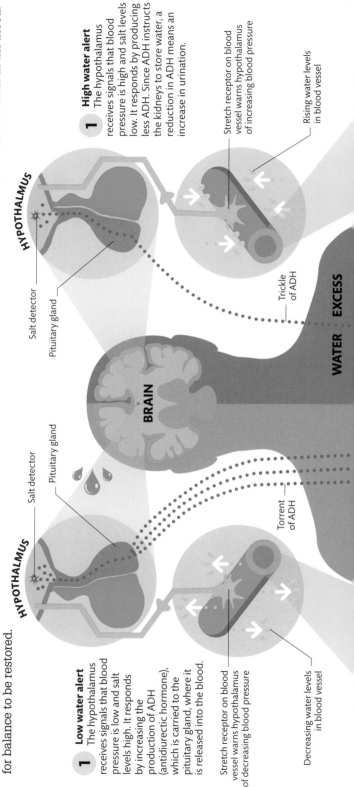

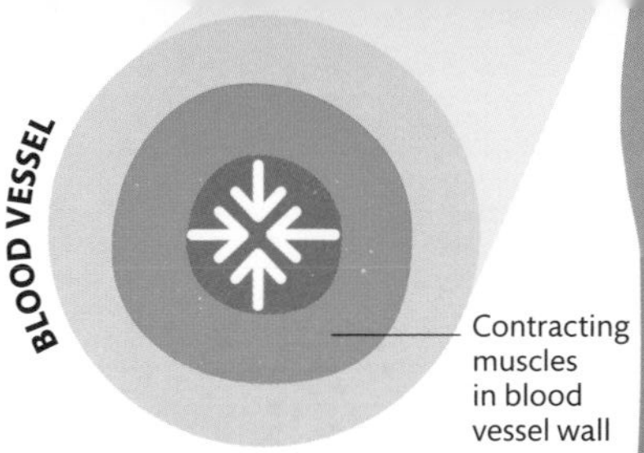

BLOOD VESSEL

Relaxing muscles in blood vessel wall

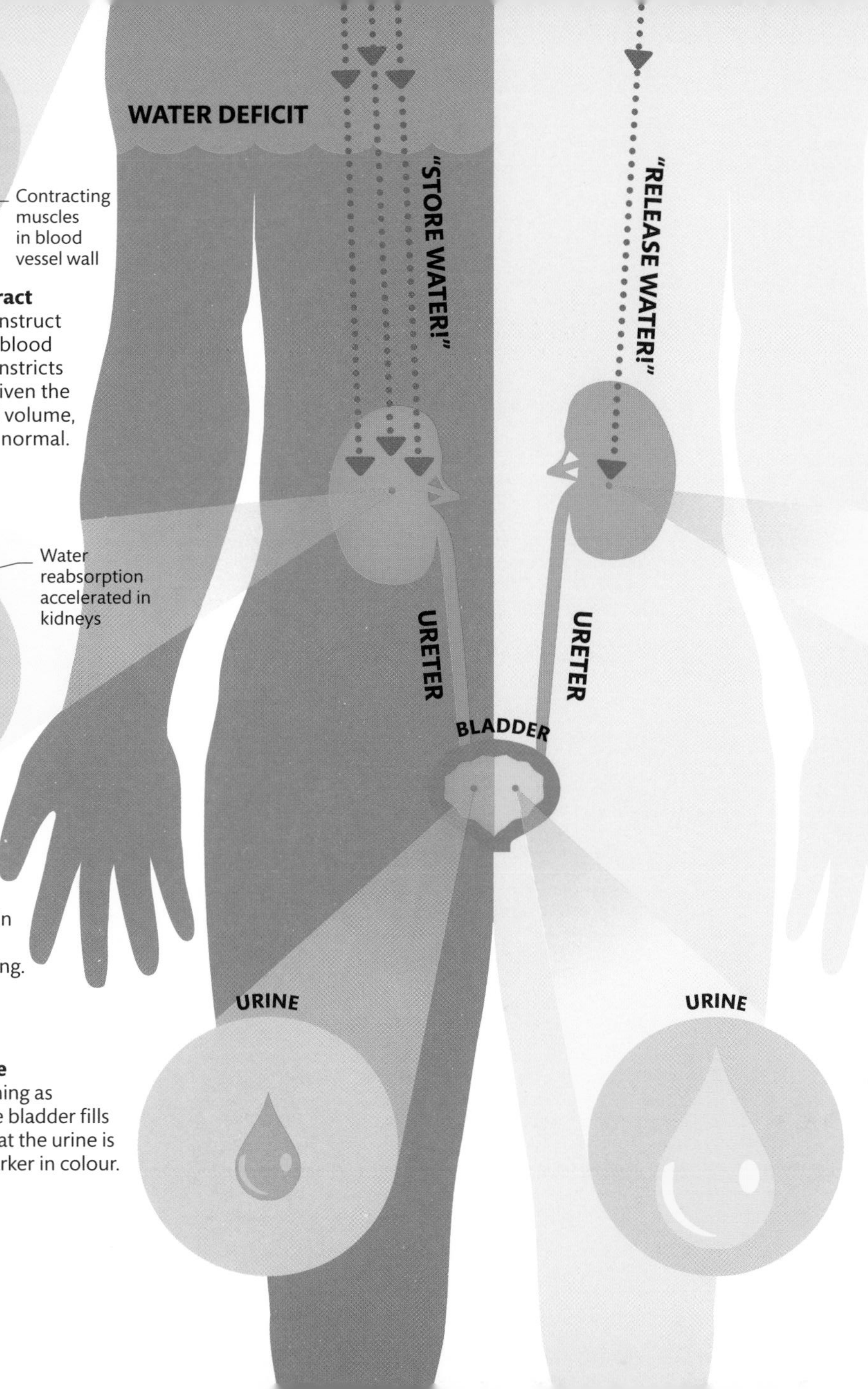

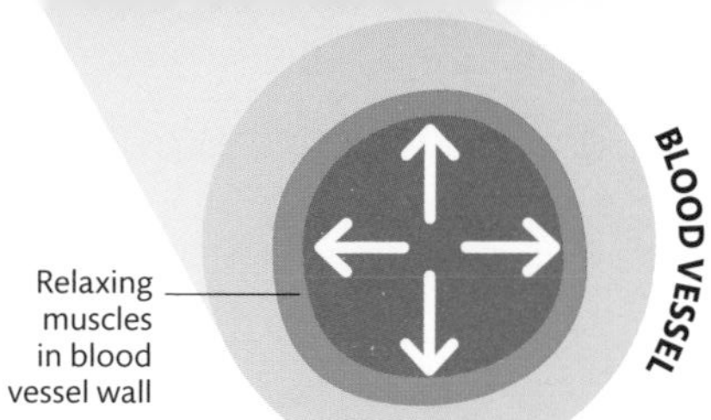

2 Blood vessels dilate
Low levels of ADH instruct the blood vessel wall muscles to relax. This expands the blood vessels and eases the blood pressure caused by the excess water.

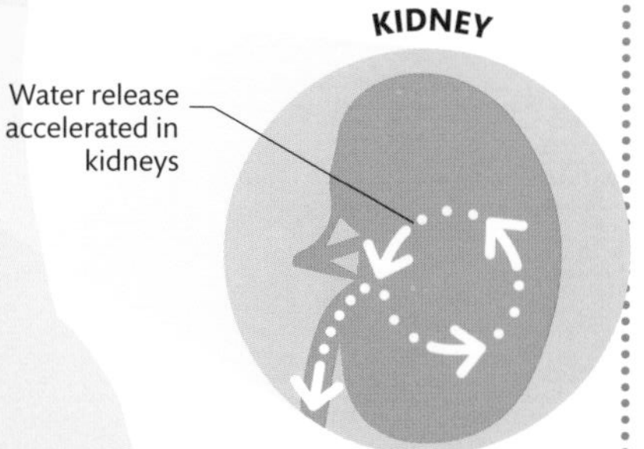

KIDNEY

Water release accelerated in kidneys

3 Water release
Low ADH levels also signal the kidneys to reduce the amount of water that they reabsorb, so that more water is added to the urine and passed out through the bladder.

4 Diluted urine
With less water being reabsorbed by the body, the bladder fills quickly and more diluted urine is produced. The more diluted the urine, the lighter its colour.

URINE

"RELEASE WATER!"

URETER

BLADDER

"STORE WATER!"

URETER

WATER DEFICIT

Contracting muscles in blood vessel wall

URINE

Water reabsorption accelerated in kidneys

KIDNEY

2 Blood vessels contract
High levels of ADH instruct muscles in the walls of the blood vessels to contract. This constricts the blood vessels, which, given the current reduction in blood volume, restores blood pressure to normal.

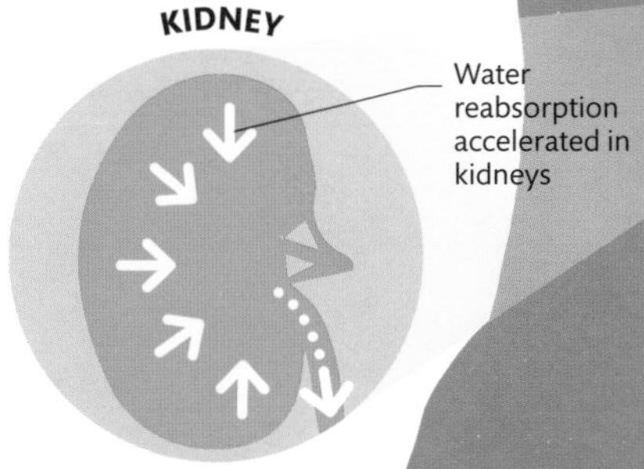

3 Water reabsorbtion
High ADH levels also signal the kidneys to reabsorb water and to retain the salts that are often lost through sweating or vomiting.

4 Concentrated urine
With the body retaining as much water as possible, the bladder fills more slowly. This means that the urine is more concentrated, and darker in colour.

BLOOD VESSEL

How the liver works

Once nutrients have entered the blood – via the mouth, stomach, and intestines – they are taken straight to the liver. Here, they are variously stored, dismantled, or turned into something new. At any one time, the liver holds some 10 percent of the body's blood supply.

Liver lobule

The liver is made up of thousands of tiny factories called lobules. Each of these contains thousands of chemical processors called hepatocytes. These do all the liver's work, albeit supported by Kupffer cells and stellate cells. Each lobule has a central outflowing vein and is six-sided, with each of its corners supporting two incoming blood supplies and an outflowing duct for bile.

Ins and outs of the liver
Blood arrives from two directions, then the liver outputs blood via the hepatic vein and bile through the bile duct.

- ·····> Blood from the intestines
- ·····> Blood from the heart
- ——> Blood to the heart
- ·····> Bile to the gallbladder

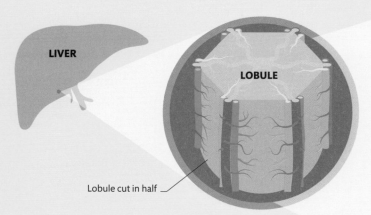

LIVER

LOBULE

Lobule cut in half

HEPATIC VEIN

HEPATIC PORTAL VENULE

HEPATIC ARTERIOLE

HEPATIC ARTERIOLE

HEPATIC PORTAL VENULE

DOUBLE BLOOD SUPPLY

An unusual fact about the liver is that it has two blood supplies. Like all other organs, it receives oxygenated blood from the heart to give it energy, but it also receives blood from the intestines, which it cleans, stores, and processes.

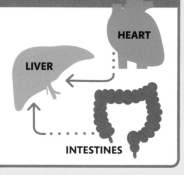

HEART

LIVER

INTESTINES

1 **Nutrients in**
Each corner of the lobule receives nutrient-rich blood from a branch of the hepatic portal vein, which comes from the intestines; this is called the hepatic portal venule. It also receives oxygen-rich blood from a branch of the hepatic artery, which comes from the heart; this is called the hepatic arteriole.

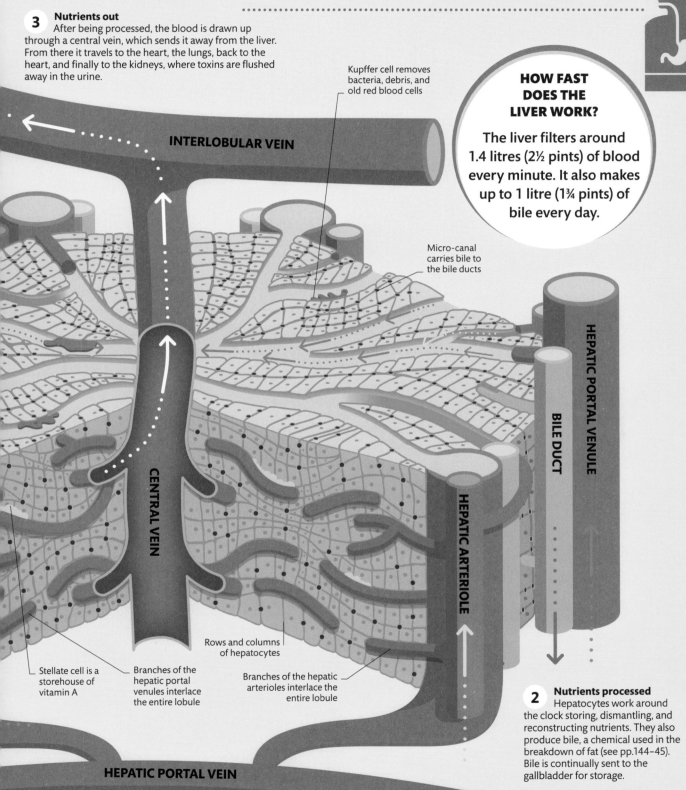

3 **Nutrients out**
After being processed, the blood is drawn up through a central vein, which sends it away from the liver. From there it travels to the heart, the lungs, back to the heart, and finally to the kidneys, where toxins are flushed away in the urine.

Kupffer cell removes bacteria, debris, and old red blood cells

HOW FAST DOES THE LIVER WORK?

The liver filters around 1.4 litres (2½ pints) of blood every minute. It also makes up to 1 litre (1¾ pints) of bile every day.

INTERLOBULAR VEIN

Micro-canal carries bile to the bile ducts

HEPATIC PORTAL VENULE

BILE DUCT

CENTRAL VEIN

HEPATIC ARTERIOLE

Stellate cell is a storehouse of vitamin A

Branches of the hepatic portal venules interlace the entire lobule

Rows and columns of hepatocytes

Branches of the hepatic arterioles interlace the entire lobule

2 **Nutrients processed**
Hepatocytes work around the clock storing, dismantling, and reconstructing nutrients. They also produce bile, a chemical used in the breakdown of fat (see pp.144–45). Bile is continually sent to the gallbladder for storage.

HEPATIC PORTAL VEIN

What the liver does

The liver is perhaps best understood as a factory – a processing plant with three main departments; processing, manufacturing, and storage. Its raw materials are the nutrients absorbed by the blood during digestion – but which department they go to depends on the body's priorities.

WHAT ELSE DOES THE LIVER DO?

It produces blood clotting proteins, which ensure that we stop bleeding when injured. People with unhealthy livers tend to bleed easily.

Glucose from carbohydrates

In a process called gluconeogenesis, the liver makes glucose out of carbohydrates when the body is low on energy.

Metabolising fat

Excess carbohydrates and proteins are converted into fatty acids and released into the bloodstream for energy. This becomes vital when glucose is running out.

Processing

The liver spends most of its time processing nutrients. This involves making sure that the right nutrients are sent to the right parts of the body, and that back-ups are provided when needed. Crucially, this also means flushing out toxic subtances.

Detoxifying the blood

Pollutants, bacterial toxins, and defensive chemicals from plants are turned into less dangerous compounds, then sent to the kidneys to be flushed out of the body.

THE REGENERATING ORGAN

Unlike other organs, which create scar tissue at sites of injury, the liver creates brand new cells when it needs them. This is lucky, since it is constantly being bombarded by unhealthy, toxic chemicals. These chemicals – which include some legal medications – frequently damage the liver, but it holds its ground by regenerating itself. Incredibly, it can lose 75 per cent of its mass and still regrow completely – all in a matter of weeks.

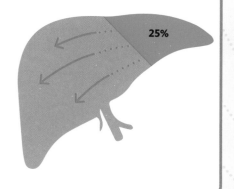

25%

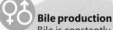 **Bile production**
Bile is constantly being produced by the liver and sent to the gallbladder for storage. It is made out of haemoglobin, which is released during the breakdown of old red blood cells.

Hormone production
The liver secretes at least three hormones, making it a key player in the endocrine system (see pp.190–91). The liver's hormones stimulate cell growth, encourage bone marrow production, and aid blood pressure control.

Manufacturing
The liver is a major manufacturing hub, turning simple nutrients into, among other things, chemical messengers (hormones), body tissue components (proteins), and a vital digestive fluid (bile). Since it is always busy, the liver also produces another precious commodity – an enormous amount of heat.

 Protein synthesis
The liver produces many proteins that are then secreted into the blood. It does so particularly when certain amino acids (the building blocks of proteins) are missing from the diet.

Storage
A great deal of stockpiling goes on in the liver, mainly of vitamins, minerals, and glycogen (the stored form of glucose). This enables the body to survive without food for days and weeks on end, and ensures that any shortfall in dietary nutrients can quickly be corrected.

Vitamins
The liver can store up to 2 years' worth of vitamin A, which is vital to the immune system. Vitamins B12, D, E, and K are also stored until needed.

THE LIVER PERFORMS SOME 500 CHEMICAL FUNCTIONS IN TOTAL

Minerals
Two vital minerals are stored in the liver: iron, which carries oxygen through our bodies; and copper, which keeps the immune system healthy. Copper is also used to make red blood cells.

Glycogen
Energy is stored as glycogen in the liver. When the body runs out of energy (see pp.158–59), the liver converts it to glucose and releases it into the bloodstream.

LIVER DAMAGE

Uniquely among the body's organs, the liver regenerates itself. However, repeated exposure to damaging agents, such as alcohol, a drug, or a virus, can eventually injure the liver. This happens when it is inundated by toxins and never gets a chance to regenerate. In this strung-out state, the liver is finally scarred – a condition known as cirrhosis. A common cause of cirrhosis is drinking too much alcohol.

Energy balance

Most of the body's cells use glucose or fatty acids for energy. To maintain a regular supply of these, the body alternates between absorbing energy (by eating) and releasing it (after which we feel hungry again). In ideal conditions, this cycle repeats itself every few hours.

Filling the tanks

Glucose and fatty acids enter our bodies through the food we eat. As blood glucose levels rise, the pancreas releases the hormone insulin. This tells muscle, fat, and liver cells to absorb and store the glucose and fatty acids as energy for the future.

Food rich in sugar

DOES FAT MAKE YOU FAT?

Only when eaten with sugary foods or carbohydrates. These foods contain glucose, which signals the body to store nutrients, and so to put on weight.

Numerous sugar molecules indicate high blood sugar level after meal

Fatty acid molecule

Glucose molecule

Fatty acids being stored in a fat cell

3 **Excess glucose stored**
Most fatty acids are stored in fat cells, which serve as resevoirs of energy. These cells also absorb excess glucose and convert it into fatty acid molecules.

Excess glucose heading for storage in a fat cell

ABSORB!

2 **Muscle burns glucose**
Muscle cells, among others, convert glucose into energy for contracting. Muscle cells also absorb fatty acids. They burn the fatty acids when glucose levels are low.

Glucose being absorbed by a muscle cell

Fatty acid being absorbed by a muscle cell

ABSORB!

1 **"Absorb!" signal sent**
After a meal, the pancreas detects high levels of sugar in the blood. In response, it releases insulin, which circulates in the blood. This readies the body's cells to open and receive nutrients. Chief among these is glucose, which all cells use for energy.

PANCREAS

Burning the fuel

As the body's cells absorb nutrients, blood glucose levels start to fall. Unless more food is digested, these levels drop to a point where the body burns fat instead of glucose for energy. Once again, this process is organized by the pancreas.

Sparse sugar molecules indicate low blood sugar level

Fatty acids being burnt in a muscle cell

3 **Muscle cell burns fat**
Here, a muscle cell receives fatty acids from a fat cell and breaks them down for energy.

Fatty acids released into the bloodstream

2 **Fat sent to muscle**
Glucagon also tells fat cells to release their stored fatty acids into the bloodstream. These fatty acids can then be used as a source of energy by other cells.

BURN!

BURN!

1 **"Burn!" signal sent**
A few hours after eating, specialized cells in the pancreas detect a drop in blood glucose levels. The pancreas releases the hormone glucagon into the bloodstream. This signals the liver to release the glucose it has stored in the form of glycogen into the bloodstream (see pp.154–55).

PANCREAS

ENERGY SUPPLY AND DEMAND

Food energy is measured in calories. A steak contains around 500 calories, as does a large packet of crisps, or 10 apples. A person at rest needs around 1,800 calories a day to maintain weight – more energy in or out tips the scales.

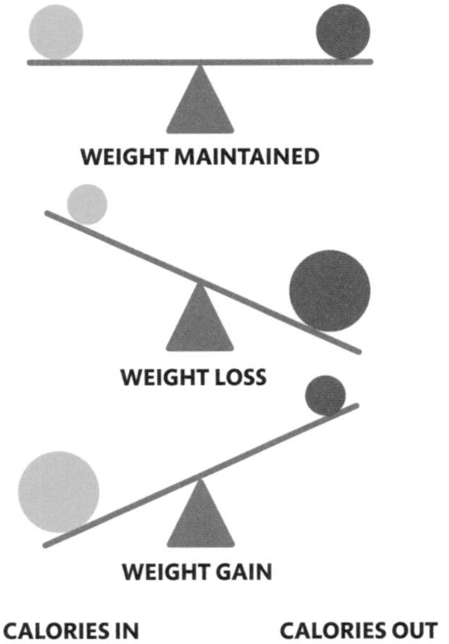

WEIGHT MAINTAINED

WEIGHT LOSS

WEIGHT GAIN

CALORIES IN   CALORIES OUT

The sugar trap

Calories are equal in terms of the amount of energy they contain, but where they come from – fat, protein, or carbohydrate – determines how they are used by the body. Some foods give us a steady source of energy, others can take us on a hormone roller coaster ride.

ARE CALORIES BAD FOR YOU?

A calorie is the amount of energy your body will gain from eating the food that contains it, so no – we need energy to live! But, if you eat too many calories, your body will store the excess as fat.

Lingering insulin

Foods that are quickly turned into sugars cause a spike in blood glucose levels (see p.158). Insulin spikes in response, causing glucose levels to plummet. The sugar crash leaves us tired and craving more sugar, while insulin lingers in our blood and prevents us from burning fat.

Rise and fall
The peak and crash of glucose and steady rise and fall of insulin levels in the blood is traced along mealtimes during a morning.

→ Glucose

→ Insulin

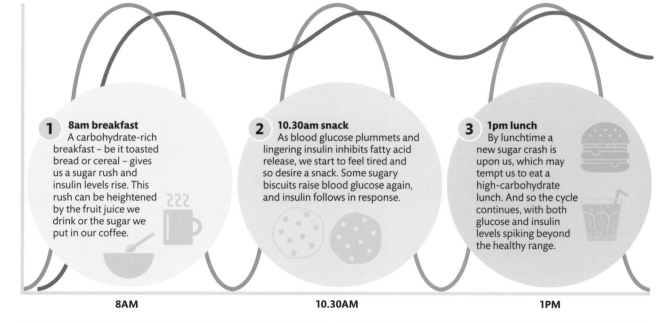

1 8am breakfast
A carbohydrate-rich breakfast – be it toasted bread or cereal – gives us a sugar rush and insulin levels rise. This rush can be heightened by the fruit juice we drink or the sugar we put in our coffee.

2 10.30am snack
As blood glucose plummets and lingering insulin inhibits fatty acid release, we start to feel tired and so desire a snack. Some sugary biscuits raise blood glucose again, and insulin follows in response.

3 1pm lunch
By lunchtime a new sugar crash is upon us, which may tempt us to eat a high-carbohydrate lunch. And so the cycle continues, with both glucose and insulin levels spiking beyond the healthy range.

8AM   10.30AM   1PM

Putting on the pounds

The sugar trap quickly leads to weight gain, and being overweight can have serious health implications. These include insulin sensitivity, insulin resistance, type-2 diabetes (see pp.201), heart disease, some types of cancer, and stroke. To avoid obesity, it is vital to keep insulin levels low, and one way of doing that is through a low-carbohydrate diet.

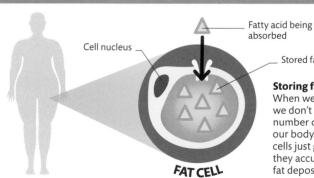

Cell nucleus

Fatty acid being absorbed

Stored fatty acid

Storing fat
When we put on fat we don't increase the number of fat cells in our body. The same fat cells just get bigger as they accumulate more fat deposits.

FAT CELL

HIGH-PROTEIN DIET

To cut out carbohydrates, some diet promoters recommend getting calories from protein and healthy fats instead. You can follow a diet in phases designed to train your body to start burning fat and rely less on carbohydrates.

Low-carbohydrate diets

A popular, if controversial, way out of the sugar trap is to limit our consumption of carbohydrates, which are otherwise broken down into sugars and stored as fat. By doing so we avoid the glucose-insulin roller coaster that ends in sugar cravings and increased fat storage. Keeping sugar and insulin levels within a healthy range enables fat, rather than glucose, to be used as an energy source.

SUGAR IS NOW THOUGHT TO BE MORE ADDICTIVE THAN COCAINE

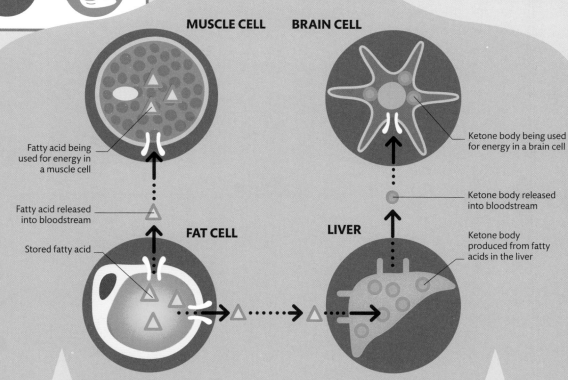

MUSCLE CELL   **BRAIN CELL**

Fatty acid being used for energy in a muscle cell

Ketone body being used for energy in a brain cell

Fatty acid released into bloodstream

Ketone body released into bloodstream

Stored fatty acid

FAT CELL   **LIVER**

Ketone body produced from fatty acids in the liver

Releasing fatty acids
When blood glucose is maintained at a healthy level, insulin levels remain low. This allows the release of fatty acids from fat cells – a process that is otherwise inhibited by insulin.

Producing ketone bodies
Unlike other tissues, the brain can't use fatty acids as an energy source. So when blood glucose is low, the liver begins to convert fatty acids into ketone bodies – molecules that provide energy for brain cells.

Feast or fast?

Two of today's most popular diets don't involve calorie counting at all. Paleolithic diets aim for an ancestral way of eating, removing the highly processed foods of today. Intermittent fasting, on the other hand, takes a more feast and fast approach, restricting when you eat rather than what you eat.

Back to basics

The theory behind palaeolithic diets is that our bodies have not evolved to consume the highly processed, sugary, carbohydrate-rich foods that are abundant in supermarkets today. The diet promotes foods that are thought to have been available to our hunter-gatherer ancestors, who lived before the advent of farming, 10,000 years ago – although the lifestyle doesn't involve reverting back to cave life. Dieters used to getting their calcium from dairy foods need to find calcium-rich alternatives, or they put themselves in danger of calcium deficiency.

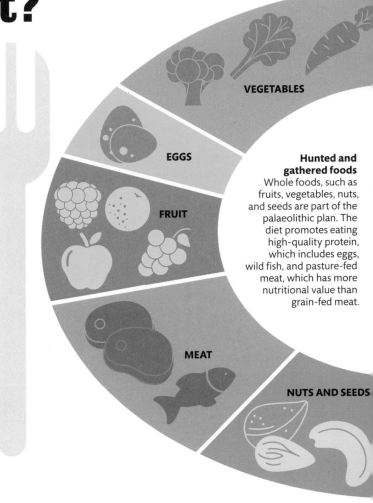

VEGETABLES

EGGS

FRUIT

MEAT

NUTS AND SEEDS

Hunted and gathered foods
Whole foods, such as fruits, vegetables, nuts, and seeds are part of the palaeolithic plan. The diet promotes eating high-quality protein, which includes eggs, wild fish, and pasture-fed meat, which has more nutritional value than grain-fed meat.

Intermittent fasting

The idea behind intermittent fasting is to take regular breaks from eating, during which the body gets all its energy from stored fat, but not for so long that it starts to break down muscle protein for energy. There are two main intermittent fasting methods; the 16:8 and the 5:2.

The 16:8 method
Followers of this regime eat during an 8-hour period every day (say noon to 8pm). The other 16 hours you fast, but luckily a lot of this time is spent sleeping, which makes it more manageable.

Key: ▓ Eating ▓ Fasting

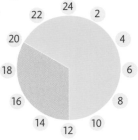

| MONDAY | TUESDAY | WEDNESDAY | THURSDAY |
| FRIDAY | SATURDAY | SUNDAY | Fasting days |

The 5:2 method
This regime restricts your daily energy intake to about 500 Calories (about one meal) per day for two days of the week. You can eat as much as you like (within reason) for the other five days of the week.

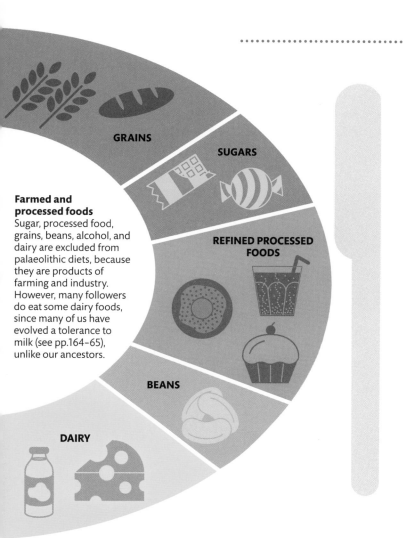

GRAINS

SUGARS

REFINED PROCESSED FOODS

BEANS

DAIRY

Farmed and processed foods

Sugar, processed food, grains, beans, alcohol, and dairy are excluded from palaeolithic diets, because they are products of farming and industry. However, many followers do eat some dairy foods, since many of us have evolved a tolerance to milk (see pp.164–65), unlike our ancestors.

ONE-THIRD OF THE WORLD'S ADULTS NOW PRODUCE THE **ENZYME** THAT DIGESTS **DAIRY SUGAR**

The glycemic index

The glycemic index (GI) is a measure of how quickly carbohydrate-containing foods increase glucose levels in the blood. The lower a food's GI value, the less it affects blood sugar levels. An attraction of palaeolithic diets is that they focus on low GI foods.

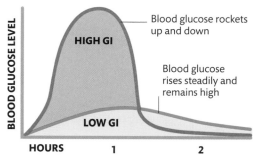

BLOOD GLUCOSE LEVEL

HIGH GI

LOW GI

Blood glucose rockets up and down

Blood glucose rises steadily and remains high

HOURS 1 2

Blood glucose levels
High GI foods rapidly increase blood sugar levels, but this is followed by a rapid decrease, leaving us feeling hungry. Low GI foods gradually increase blood sugar levels, leaving us feeling full for longer.

Natural fat-burning

Exercising when your body is naturally burning fat may give your workout more punch. A run before breakfast, for example, takes advantage of the fact that your body is already burning fat after fasting all night. A run in the evening, however, is more likely to be fuelled by blood glucose supplied by the day's food. For this reason, morning exercise is generally more effective for losing weight.

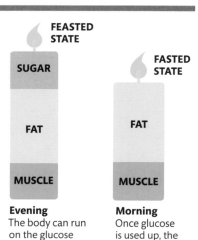

FEASTED STATE

SUGAR

FAT

MUSCLE

Evening
The body can run on the glucose from a meal for about 3–5 hours.

FASTED STATE

FAT

MUSCLE

Morning
Once glucose is used up, the body starts to burn fat stores.

BRAIN HEALTH

There is evidence that fasting improves brain health. Intermittent fasting in particular puts the neurons under mild stress – much like our muscles are stressed by exercise. This stress causes the release of chemicals that help in the growth and maintenance of neurons.

FASTED BRAIN

Neuron

Digestive problems

Digestive problems can range from temporary discomfort after eating to life-long persistent disorders. In most cases, the treatment is simply to avoid the foods that cause the symptoms.

Lactose intolerance

Many adults lack the enzyme lactase, which is needed to break down lactose, the sugar found in milk. All healthy babies have it, but most of us stop producing this enzyme after weaning. Only about 35 per cent of the world's population have acquired a mutation that allows them to produce lactase into adulthood.

WHO ISN'T LACTOSE INTOLERANT?

Countries that have a long history of dairy farming tend to have populations that have adapted to drinking milk into adulthood. Most of these countries are in Europe.

Lactose

Lactase enzyme

2 Lactose digested by lactase
Lactase breaks lactose into two smaller sugars – galactose and glucose.

SMALL INTESTINE

Glucose

1 Lactose in small intestine
When the cells that line the walls of the small intestine encounter the sugar lactose, they start to produce the digestive enzyme lactase.

Galactose

3 Galactose and glucose absorbed
These two smaller sugars are then absorbed into the bloodstream by the small intestine.

2 Bacterial fermentation
Bacteria living in the large intestine (see pp.148–49) ferment the lactose, producing gas and acids in the process.

3 Disruption in the bowl
The gas produced by fermentation causes bloating and discomfort, while the acids draw water into the bowel, leading to diarrhoea.

Gas and acids released by bacteria

LARGE INTESTINE

Undigested lactose enters the large intestine

1 Undigested lactose
If lactase isn't present, then lactose can't be absorbed and instead passes into the large intestine.

Bacteria fermenting lactose

BRINGING IT UP

One way the body avoids digestive problems is by vomiting. When we eat something rotten or poisonous, the stomach, the diaphragm, and the abdominal muscles all contract, forcing the food back up through the oesophagus and out through the mouth.

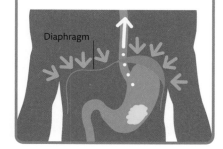

Diaphragm

Irritable bowel syndrome

IBS is a long-term condition that can cause stomach cramps, bloating, diarrhoea, and constipation. It is poorly understood, but seems to be triggered by stress, lifestyle, and certain types of food.

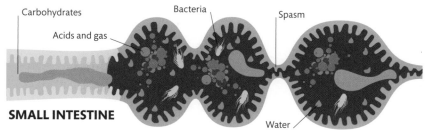

Carbohydrates
Acids and gas
Bacteria
Spasm
Water
SMALL INTESTINE
LARGE INTESTINE

1 Bacterial fermentation
Carbohydrates that are poorly absorbed may increase the amount of water in the intestinal tract. Once in the large intestine, these carbohydrates are fermented by bacteria, producing acids and gas.

2 Bowel spasms
IBS causes bowel spasms, which can block the waste and gas from passing through. Alternatively, it can cause the waste to move too quickly, preventing water reabsorption and causing diarrhoea.

Gluten intolerance

Many people experience abdominal pain, fatigue, headaches, and even numbness of the limbs when they eat gluten – a protein found in grains such as wheat, barley, and rye. These symptoms are indicators of various gluten-related disorders, ranging from gluten sensitivity to coeliac disease.

RYE BREAD

BEER

PASTA

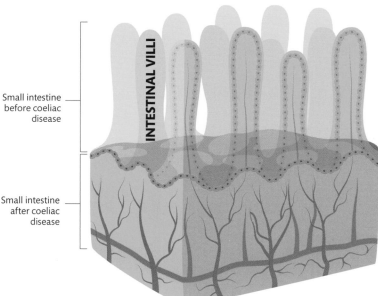

INTESTINAL VILLI

Small intestine before coeliac disease

Small intestine after coeliac disease

Gluten sensitivity
Lethargy, mental fatigue, cramps, and diaorrhea are all symptoms of gluten sensitivity, which is only cured by avoiding all gluten products – including rye bread, beer, and pasta. Gluten sensitivity does not damage the intestines like coeliac disease does.

Coeliac disease
Coeliac disease is a serious genetic disorder that causes the body's immune system to attack itself when it encounters gluten. This immune response causes damage to the lining of the small intestine and so inhibits the absorption of nutrients. Left unchecked, it can totally destroy the small intestine's little fingerlike projections, or villi.

FIT AND
HEALTHY

Body battleground

Humans are attacked on a daily basis by a host of marauding invaders, for whom the body is an ideal place to feed and reproduce. Ranged against them are the body's defence forces. Any harmful microbe, or pathogen, that breaks through the outer barriers is met with a quick, local response at the site of the infection. If this doesn't work, a second team is called into action.

Invaders

Bacteria and viruses are the major causes of disease in humans. Parasitic animals, fungi, and toxins can also prompt the immune system into action. All these microbes are constantly adapting and evolving to find new ways to avoid detection and destruction by the immune system.

Fungi
Most are not dangerous, but some can be harmful to health.

Parasitic animals
Live on or inside humans and may carry other pathogens into their host.

Bacteria
Tiny, single-celled organisms taken into the body by eating, breathing, or through breaks in the skin.

Viruses
Viruses need other living cells to multiply and can lie dormant inside their host's cells for long periods.

Toxins
These are substances capable of causing disease or a reaction that could prove deadly to the human body.

Secretions
Fluids such as mucus, tears, oils, saliva, and stomach acid can trap pathogens or break them down with enzymes.

Complement proteins
As many as 30 different proteins circulate in the blood, ramping up the immune response by marking pathogens for destruction or causing them to burst.

Dendritic cells
These phagocytes (microbe eaters) engulf pathogens and play an important role in spurring B and T cells into action.

Barricades

Epithelial cells are the body's main physical defence against pathogens. The cells are tightly packed together to prevent anything penetrating them. They also secrete liquids that act as a further barrier against pathogens.

Epithelium
Epithelial cells form the skin and membranes that line all of the body's openings, such as the mouth, nose, oesophagus, and bladder.

EPITHELIUM
SECRETIONS

Front-line troops

Pathogens that break through the barriers are met with an immediate response known as the innate immune system. This is a group of cells and proteins that respond to alarm signals from damaged or infection-stressed cells. Some target and mark invading organisms for destruction, while others (phagocytes) eat up the pathogens.

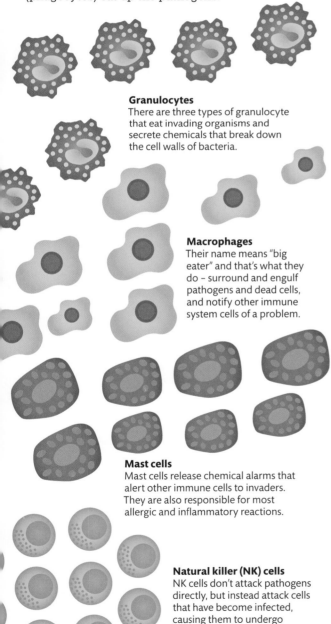

Granulocytes
There are three types of granulocyte that eat invading organisms and secrete chemicals that break down the cell walls of bacteria.

Macrophages
Their name means "big eater" and that's what they do – surround and engulf pathogens and dead cells, and notify other immune system cells of a problem.

Mast cells
Mast cells release chemical alarms that alert other immune cells to invaders. They are also responsible for most allergic and inflammatory reactions.

Natural killer (NK) cells
NK cells don't attack pathogens directly, but instead attack cells that have become infected, causing them to undergo apoptosis (see p.15).

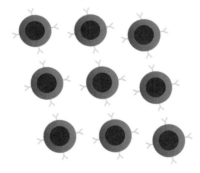

HOW MANY INFECTIOUS DISEASES CAN THE IMMUNE SYSTEM RESPOND TO?

It is thought that B cells alone can produce enough different antibodies to deal with 1 billion different types of pathogen.

Killer cavalry

If the front-line response hasn't contained the infection within 12 hours, the adaptive immune system swings into action. This system remembers previous exposures to the pathogen to launch a specific, targeted response.

B cells
B cells are a special type of cell that can be trained to produce antibodies in response to the presence of a particular pathogen. They can multiply rapidly to increase the response.

Antibodies
Antibodies are Y-shaped proteins produced by B cells. They stick to the surface of invaders and mark them out for destruction by phagocytes.

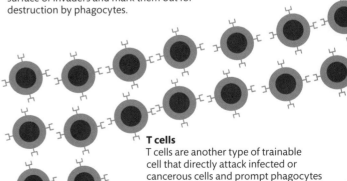

T cells
T cells are another type of trainable cell that directly attack infected or cancerous cells and prompt phagocytes to eat pathogens. Some T cells also stimulate B cells to produce antibodies.

Friend or foe?

The immune system has to distinguish the harmful pathogens that invade our body from the body's own cells and friendly microbes – in other words, recognize friends from foes. The body puts its most potent immune cells – B and T cells – through safety checks to prevent them from attacking us.

Self and nonself

Every cell in the body is coated in groups of molecules that are unique to each individual. The main function of these molecules is to display fragments of protein made by the body and friendly microbes so that the immune system learns to tolerate them and recognize them as "self".

Antigens, specific to each person, coat this body cell

BODY CELL

Antigen of a different shape. All antigens have a characteristic shape known as an epitope

FOREIGN CELL

Self tolerance
All body cells carry "self" surface marker proteins, or antigens, allowing them to live in harmony with other cells. If the immune system loses its ability to recognize self markers it can lead to autoimmune diseases.

Nonself markers
Foreign cells carry their own surface marker proteins, which trigger an immune response. Even the proteins you eat may be identified as foreign unless they are broken down first by the digestive system.

TRANSPLANTS

Compatibility is examined before an organ transplant is given, but if it is not a close enough match the recipient's immune system may launch an attack on the donated tissue and start to destroy it. Transplant recipients may have to take immunosuppressant drugs to try to minimize this complication.

Starting point
Both B cells (which produce antibodies to kill invaders, see pp.178–79) and T cells (which kill invaders directly, see pp.180–81) start life as stem cells in the bone marrow.

1 Bone marrow
B cells mature and are tested in the bone marrow. Any that bond with self proteins in the marrow are deactivated and killed by apoptosis (see p.15).

BONE

B cell receptor

B CELL

2 B cell
If a B cell passes the self test, it is released from the bone marrow into the lymphatic system. This is a network of vessels that runs parallel to blood vessels and carries immune cells around the body.

ONLY **2 PER CENT** OF **T CELLS PASS THEIR TRAINING** – THE REST ARE **REJECTED** AS THEY **MIGHT ATTACK US!**

DO IDENTICAL TWINS HAVE THE SAME IMMUNE SYSTEM?

No. Immunity is shaped by what each person is exposed to in life, so it is very individual.

Tested to destruction

When the T cells and B cells of the immune system are forming, they generate random receptors and put them on their surface. Because this process is random, it is possible that these receptors might bind strongly with "self", or friendly, antigens. Therefore, these cells go through vigorous testing before being released into the body. Those that bind to the body's own proteins are destroyed.

Bean-shaped lymph nodes, many of which are in the armpits and groin, are reservoirs for B cells, T cells, and other immune cells

LYMPH NODE

T cells
B cells
Other immune cells

DESTINATION
If invaders are present in the body's circulation, eventually, they have to pass the lymph nodes, where B cells and T cells lie in wait. The cells activate when they encounter an alien antigen that matches their receptors.

1 THYMUS
T cells move to the thymus (a specialized lymph gland found in front of the heart) where they mature. Their receptors are tested to make sure they don't form strong bonds with self proteins.

THYMUS

T cell receptor

T CELL

2 T CELL
Mature T cells are released into the lymph and blood. Regulatory T cells are a subtype that provide an extra check on the self-tolerance of other T cells.

Compatibility

Compatibility tests look at the likelihood of a recipient's immune system attacking donated tissue. Red blood cells carry extra self markers called blood groups. Two of them, the ABO and Rhesus groups, prompt an immune reaction to donated blood from a different group. People with blood group O, for example, will launch a response to blood from any other group because they carry both anti-A and anti-B antibodies.

Blood group A
The red blood cells display A antigens on their surface and antibodies to B antigens are found in the blood plasma.

A antigen   Anti-B antibody

Blood group B
The red blood cells display B antigens on their surface and the plasma has antibodies to A antigens.

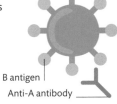

B antigen
Anti-A antibody

Blood group AB
The red blood cells display both A and B antigens on their surface, but there are no antibodies in the blood plasma.

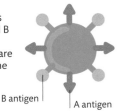

B antigen   A antigen

Blood group O
The red blood cells display neither A nor B antigens on their surface, but the blood plasma carries both types of antibody.

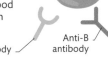

Anti-A antibody   Anti-B antibody

Germs are us

The microbes that live peacefully in and on our body are a big part of staying healthy. These microbes – mostly bacteria and fungi – have benefits that range from keeping our skin healthy by eating dead cells to helping us digest food.

Your local neighbourhood

Just as towns may be built around a particular resource, microbes collect around specific areas of the body. On the skin, for example, they are most abundant around sweat glands and hair follicles where they are more likely to find the nutrients they need to survive. The conditions in each area of the body – moist, dry, acidic – also determines which species can live there. Skin has the greatest diversity of microbes. Those on the oily back are different to those on the drier front.

NOSE

MOUTH

Microbes are carried through the air, adding to the resident population of microbes living in the nose

At least 600 different species of microbe live in the mouth

Bacteria migrate into the mammary glands from the skin and can be passed to a baby in milk

MAMMARY GLAND

The forearm has more species than any other area of the skin because of its frequent contact with objects

FOREARM

Friendly microbes produce chemicals that suppress the growth of harmful pathogens in the genital regions of men and women

BELLY BUTTON

GUT

GENITALS

It's bacteria that put the O into BO – they feed on sweat and turn it smelly

ARMPIT

The navel is home to unusual species that enjoy the dry, oilless habitat

The gut contains a relatively low diversity of species, but by far the greatest in quantity

The community here changes with everything we touch and every time we wash

HAND

AM I A HABITAT FOR RARE WILDLIFE?

Quite possibly. In a study of 90 belly buttons, researchers found 1,400 species of bacteria that had never been found on human bodies before, some of them new to science.

What's living where

The graphic shows the main types of organism found in or on regions of the body. Large icons indicate species that comprise more than 50 per cent of the population.

Bacteria

Bacteriodetes
Proteobacteria
Staphylococcaceae
Firmicutes
Corynebacteria
Actinobacteria

Fungi

Malassezia
Candida
Aspergillus
Other fungi

Viruses

Living on bacteria
Living in our cells

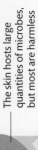

The skin hosts large quantities of microbes, but most are harmless

Naturally moist hotspots are dominated by species that thrive in warm, wet conditions

SKIN

BACK OF KNEE

MICROBIAL CELLS OUTNUMBER HUMAN CELLS BY 10 TO 1

Feet are dominated by fungi – around 100 species thrive in their cool and damp environment

SOLES OF FEET

Beneficial microbes

Science is still revealing the different species that live within the human microbiome, let alone their many benefits. Some benefits are direct, such as eating dead skin and changing the chemical environment to prevent harmful microbes from growing. Others are less obvious, such as the calming effect some gut bacteria have on the immune system by reducing inflammation. Medicines, such as antibiotics, can also have devastating effects, wiping out the good microbes as well as the bad.

Chemical released by bacteria prompts T cell into action

Immune cells no longer trigger inflammation

Bacterium

Epithelial cell

T cell releases inhibitors

Happy bacteria = healthy gut
Eating the right foods helps good bacteria to thrive. They produce chemicals that damp down inflammation in the gut, which would allow bad bacteria to penetrate the epithelial wall.

Birthday presents

Babies start to build their own microbiome at birth by picking up some of their mother's microbes as they pass through the birth canal. These bacteria start to produce chemicals that encourage other beneficial microbes to colonize. Many factors can influence the development of the microbiome; different species will colonize depending on how the baby is delivered (Caesarian babies have different bacteria), whether a baby is breast fed, and who it has contact with.

ARE WE TOO CLEAN?

It's possible that our obsession with antibacterial cleansers is taking its toll on friendly microbes. Some studies have shown that excessive handwashing can lead to the growth of more harmful microbes – but this is debatable, since other studies have shown the opposite.

Damage limitation

When a physical barrier such as skin is damaged, the immune system works quickly to repair it and defend the body against infection. The local immune cells swing into action against the first invaders, calling for more specialist reinforcements if there are more than they can cope with.

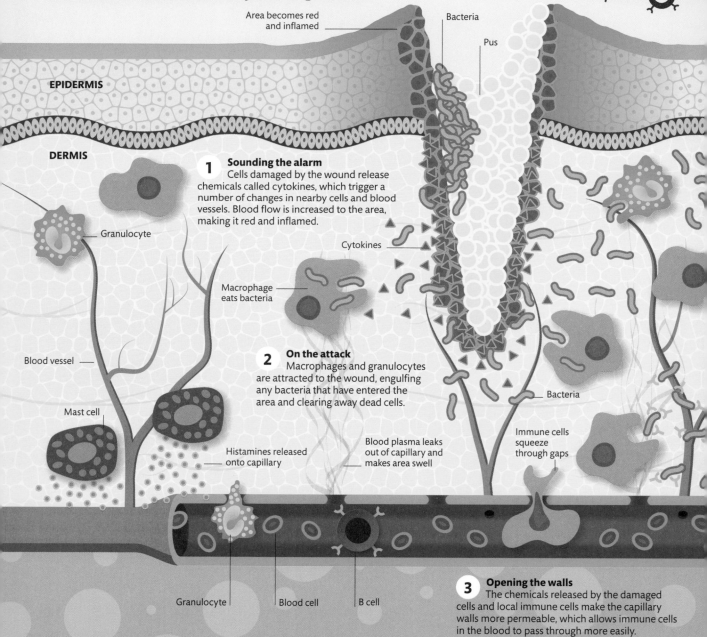

Area becomes red and inflamed

Bacteria

Pus

EPIDERMIS

DERMIS

1 Sounding the alarm
Cells damaged by the wound release chemicals called cytokines, which trigger a number of changes in nearby cells and blood vessels. Blood flow is increased to the area, making it red and inflamed.

Granulocyte

Cytokines

Macrophage eats bacteria

Blood vessel

2 On the attack
Macrophages and granulocytes are attracted to the wound, engulfing any bacteria that have entered the area and clearing away dead cells.

Bacteria

Mast cell

Histamines released onto capillary

Blood plasma leaks out of capillary and makes area swell

Immune cells squeeze through gaps

Granulocyte

Blood cell

B cell

3 Opening the walls
The chemicals released by the damaged cells and local immune cells make the capillary walls more permeable, which allows immune cells in the blood to pass through more easily.

Call to arms

A number of immune cells, such as macrophages, mast cells, and granulocytes, live in the dermis. If the skin is cut, mast cells detect the injured cells and release histamines that cause nearby blood vessels to swell. This increases blood flow to the area, making the wound feel hot, but it also brings other immune cells to the site quickly. The formation of pus is an indication that bacteria have got into the wound – pus is the accumulated remains of dead immune cells.

WHY DO CUTS TAKE LONGER TO HEAL WHEN WE'RE OLDER?

Blood vessels can become more fragile as you get older, which makes it more difficult to deliver immune cells to the wound.

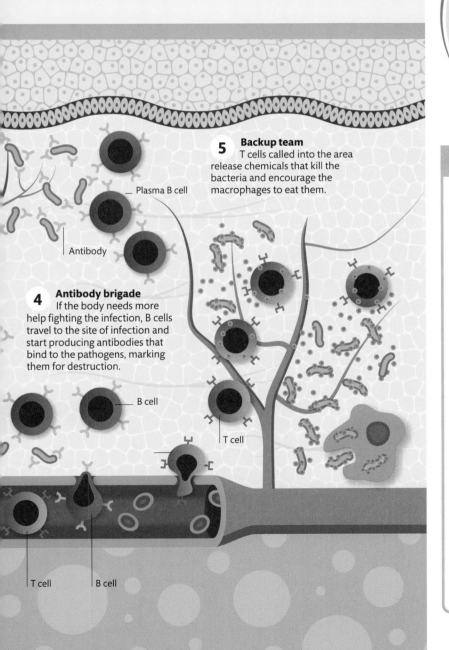

5 Backup team
T cells called into the area release chemicals that kill the bacteria and encourage the macrophages to eat them.

Plasma B cell

Antibody

4 Antibody brigade
If the body needs more help fighting the infection, B cells travel to the site of infection and start producing antibodies that bind to the pathogens, marking them for destruction.

B cell

T cell

T cell

B cell

MAGGOT THERAPY

If a wound in the skin isn't healing properly or responding to conventional treatment, maggots may be the answer. These little fly larvae are particularly precise in digesting dead cells while leaving the healthy cells alone. As they eat, the maggots secrete antimicrobial chemicals that protect the maggot but which are also effective at killing bacteria, even those resistant to antibiotics. These secretions also help inhibit inflammation of the wound, contributing to the healing process.

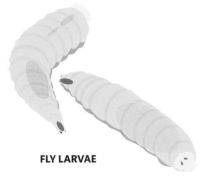

FLY LARVAE

Bacteria

Bacteria are microscopic organisms that are usually harmless, but can sometimes cause disease. Bacteria are responsible for some globally important diseases, such as tuberculosis and pneumonia.

SALMONELLA
(food poisoning)

VIBRIO
(cholera)

Flagellum

TREPONEMA
(yaws, syphilis)

STREPTOCOCCUS
(pneumonia, bronchitis)

Viruses

Viruses are the smallest and simplest organisms of all, made up of only their genetic material (DNA or RNA) in a protein coat. Unlike other pathogens, viruses need the host's cells to live and replicate.

Capsid
(protein
coat)

ADENOVIRUS
(tonsilitis, conjunctivitis)

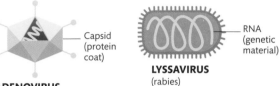

RNA
(genetic
material)

LYSSAVIRUS
(rabies)

Surface protein

Envelope

Capsid

LENTIVIRUS
(HIV/AIDS)

HERPESVIRUS
(hepatitis B, cold sores)

Antibiotics

Commonly used for bacterial infections, antibiotics break down the walls of bacteria or interrupt their growth. However, they can't distinguish the good bacteria from the bad.

Vaccination

The best way of preventing the spread of viral infections is through vaccination. A vaccine primes the immune system to recognize the virus and launch an immediate attack (see pp.184–185).

Infectious diseases

Bacteria, viruses, parasites, and fungi live in and on us all the time. Most are harmless, but certain species are pathogens – they can cause an illness if a change in conditions allows them to thrive. Other diseases are passed to us from people or animals. A fever is almost always a sign that an infection is taking hold.

Unwanted visitors

Organisms that live off the body's cells or tissues are called parasites. There are five main types: bacteria, viruses, fungi, and animals and protozoans. When they find favourable conditions they multiply rapidly but may produce harmful products or effects that make us feel ill, prompting our immune system to swing into action.

A SINGLE SNEEZE CONTAINS 100,000 GERMS

Animals and protozoans

We also face attacks from tiny animals and unicellular organisms called protozoans that live on or inside the body. Some are large enough to see with the naked eye, such as worms, or they may be microscopic, such as Giardia, the protozoan that causes diarrhoea.

Two flagellae

GIARDIA
(diarrhoea)

NEMATODE
(Guinea worm, thread worm)

Nucleus

TRICHOMONAS
(urethritis, vaginitis)

Flagellum

Fungi

Fungi are always present in and on the body, but sometimes pathogenic species take hold and cause diseases such as athlete's foot or thrush.

COCCIDIOIDES
(valley fever)

CRYPTOCOCCUS
(lung or meningeal cryptococcosis)

Arthrospores

Spore-bearing body

ASPERGILLUS
(lung infections)

Prevention

The best strategy against this type of infection is to avoid activities and areas where there are known health hazards, be wary of unsafe food and water sources, and to take recommended precautionary drugs.

Antifungal medications

Fungal infections are treated according to whether they are internal or external. The active ingredients either attack the fungus directly by breaking down its cell walls, or prevent it from growing.

How diseases spread

There are many infectious diseases but some affect relatively few individuals and are local to a small area – only diseases that spread easily by person-to-person contact are said to be contagious. Many pathogens travel between people by less direct means – through the air or in water, on objects someone has touched, or in contaminated food. Zoonotic diseases are animal infections that can spread to humans, usually through bites.

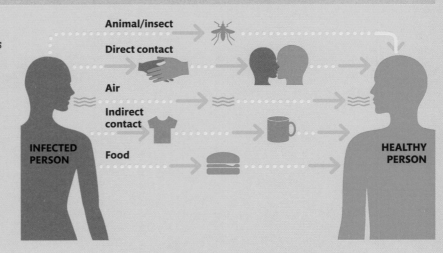

Animal/insect

Direct contact

Air

Indirect contact

Food

INFECTED PERSON

HEALTHY PERSON

Looking for trouble

If an infection becomes too great for the initial immune system to deal with, a second, more targeted force springs into action. B cells learn to recognize harmful microbes that have attacked the body in the past. They can then produce antibodies that will surround the pathogen and tag it for destruction by other immune cells.

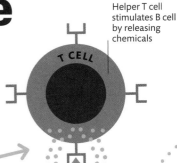

Helper T cell stimulates B cell by releasing chemicals

T CELL

B CELL

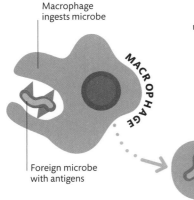

Macrophage ingests microbe

Foreign microbe with antigens

MACROPHAGE

Macrophage puts the antigens on its outer membrane, then presents them to a B cell and a helper T cell

Microbe is digested and is broken into pieces

B cell duplicates to produce two types of clone – memory B cells and plasma cells

1 **Presenting antigens**
When a macrophage ingests a pathogenic microbe it breaks it up and puts the microbe's antigens (surface proteins) onto its cell wall. This is known as an antigen-presenting cell.

2 **Helping hand**
The B cell starts to get ready when it binds to an antigen, but it isn't fully activated until a helper T cell recognizes and binds to that same antigen. The helper cell then releases chemicals that prompt the B cell to produce antibodies.

Activating antibodies

B cells are a type of white blood cell that constantly patrol the blood vessels or lie waiting in the lymph nodes (see pp.170–171). When a B cell encounters an antigen it recognizes, it becomes primed and ready to clone itself. This can only happen when another cell of the immune system, the helper T cell, recognizes and binds to that same antigen, triggering the B cell to clone itself and release antibodies.

A SINGLE **B CELL** MAY HAVE **UP TO 100,000 ANTIBODIES** ON ITS OUTER SURFACE

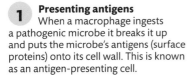

TESTING FOR ANTIBODIES

Blood tests show the levels of immunoglobulins (another name for antibodies) present during infections. IgM is a large antibody that the body produces at the first sign of infection, but it quickly disappears. IgG is a more specific, lifelong antibody that is produced during a later infection. A high IgM value shows you have a current infection, whereas IgG simply means you have been infected by a pathogen in the past.

The IgM complex has five times as many antibodies available to deal with pathogens than IgG

IgG

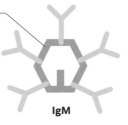

IgM

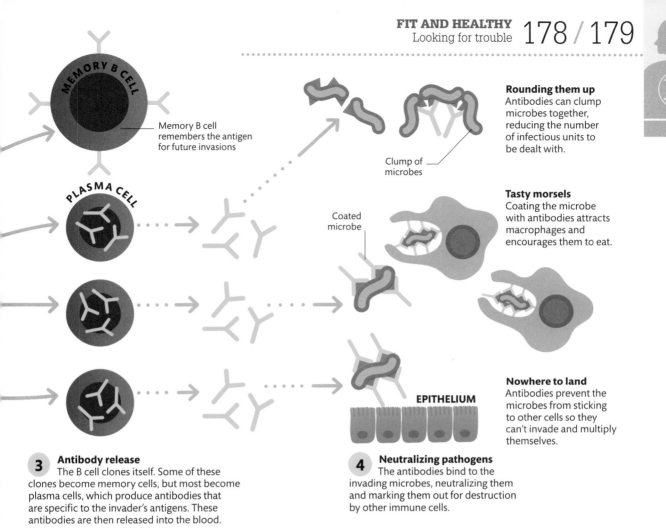

Rounding them up
Antibodies can clump microbes together, reducing the number of infectious units to be dealt with.

Clump of microbes

Tasty morsels
Coating the microbe with antibodies attracts macrophages and encourages them to eat.

Coated microbe

Nowhere to land
Antibodies prevent the microbes from sticking to other cells so they can't invade and multiply themselves.

EPITHELIUM

MEMORY B CELL

Memory B cell remembers the antigen for future invasions

PLASMA CELL

3 **Antibody release**
The B cell clones itself. Some of these clones become memory cells, but most become plasma cells, which produce antibodies that are specific to the invader's antigens. These antibodies are then released into the blood.

4 **Neutralizing pathogens**
The antibodies bind to the invading microbes, neutralizing them and marking them out for destruction by other immune cells.

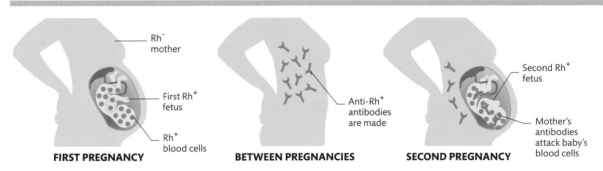

Rh⁻ mother

First Rh⁺ fetus

Rh⁺ blood cells

FIRST PREGNANCY

Anti-Rh⁺ antibodies are made

BETWEEN PREGNANCIES

Second Rh⁺ fetus

Mother's antibodies attack baby's blood cells

SECOND PREGNANCY

Rhesus babies

The Rhesus factor (Rh) is a protein on the surface of red blood cells – people who have it are called Rh+. When an Rh– mother is exposed to the blood of her Rh+ fetus (from the father's Rh+ gene) during birth, she makes antibodies against it. These antibodies may attack future Rh+ embryos, but an injection of anti-Rh+ antibodies early in the pregnancy usually reduces this danger.

Not-so-safe haven
Antibodies produced in response to the baby's blood mingling with the mother's during birth will prompt her immune system to attack the next Rh+ child she conceives. This is because her antibodies can actually cross the placenta into the baby's blood.

Assassination squad

The immune system can prime some cells to go out into the body and attack the invasion one-on-one. These are known as T cells. They hunt down infected and abnormal cells, then destroy them.

Keeping control

T cells are a type of white blood cell that play a key role in dealing with infections. Circulating in the blood and lymph, the T cells look for foreign antigens on the surface of body cells. These characteristic proteins show that the cells have been invaded by a microbe or that they have developed a dangerous abnormality. T cells also marshal the actions other immune cells and prime B cells to produce antibodies.

REGULATOR T CELLS ARE VITAL IN PREVENTING AUTOIMMUNE DISEASES

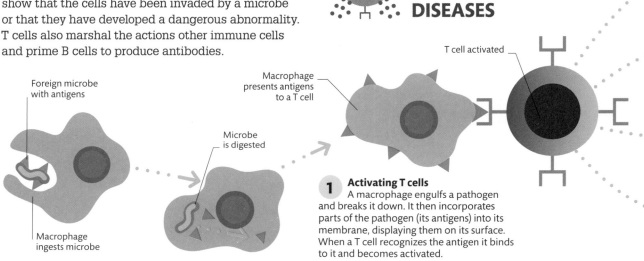

Foreign microbe with antigens

Macrophage ingests microbe

Microbe is digested

Macrophage presents antigens to a T cell

T cell activated

1 Activating T cells
A macrophage engulfs a pathogen and breaks it down. It then incorporates parts of the pathogen (its antigens) into its membrane, displaying them on its surface. When a T cell recognizes the antigen it binds to it and becomes activated.

Cornering cancer

Immunotherapy is a treatment designed to help the immune system fight cancer. There are many different ways of doing this. All of them either make the cancer cells more easily identified by the immune system or boost the immune system by multiplying cells or cytokines in the lab before injecting them back into the patient.

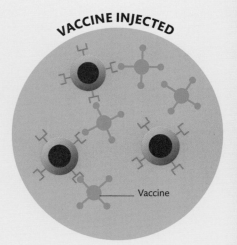

NO RESPONSE

VACCINE INJECTED

Cancer cell

T cell

Vaccine

Cancer vaccines
Vaccines form one of the methods of immunotherapy being developed. They prompt the immune system to target only cancerous cells.

1 No threat
Cancer is the uncontrolled division of abnormal cells. The immune system may not recognize these cells as abnormal because they are the body's own cells.

2 Identifying the adversary
Cancerous cells have "self" antigens on their surface but also produce their own antigens. A vaccine is designed to match the shape of the cancer antigen.

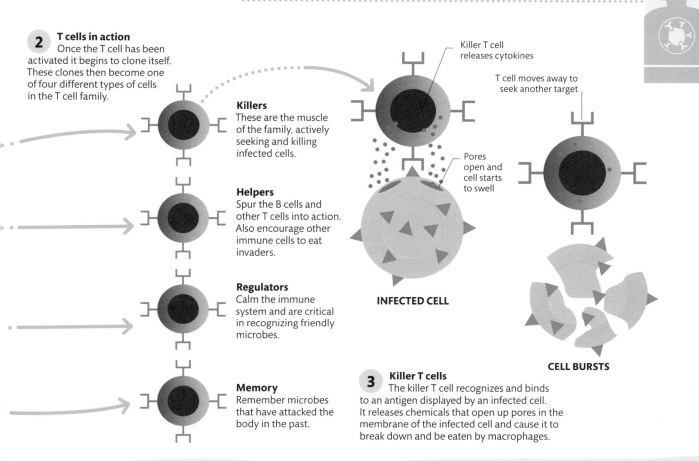

2 T cells in action
Once the T cell has been activated it begins to clone itself. These clones then become one of four different types of cells in the T cell family.

Killers
These are the muscle of the family, actively seeking and killing infected cells.

Helpers
Spur the B cells and other T cells into action. Also encourage other immune cells to eat invaders.

Regulators
Calm the immune system and are critical in recognizing friendly microbes.

Memory
Remember microbes that have attacked the body in the past.

Killer T cell releases cytokines

T cell moves away to seek another target

Pores open and cell starts to swell

INFECTED CELL

CELL BURSTS

3 Killer T cells
The killer T cell recognizes and binds to an antigen displayed by an infected cell. It releases chemicals that open up pores in the membrane of the infected cell and cause it to break down and be eaten by macrophages.

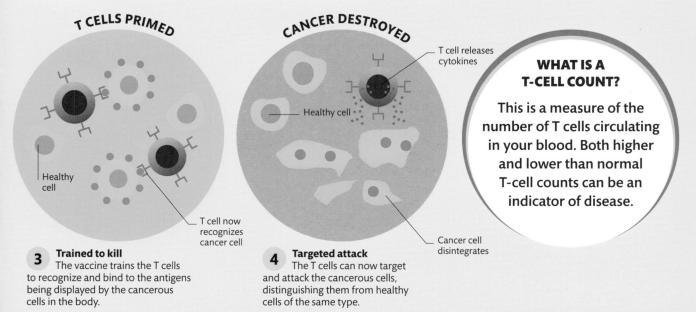

T CELLS PRIMED

Healthy cell

T cell now recognizes cancer cell

3 Trained to kill
The vaccine trains the T cells to recognize and bind to the antigens being displayed by the cancerous cells in the body.

CANCER DESTROYED

T cell releases cytokines

Healthy cell

Cancer cell disintegrates

4 Targeted attack
The T cells can now target and attack the cancerous cells, distinguishing them from healthy cells of the same type.

WHAT IS A T-CELL COUNT?

This is a measure of the number of T cells circulating in your blood. Both higher and lower than normal T-cell counts can be an indicator of disease.

Cold and flu

The reason why you are ailed by colds again and again is because the virus mutates each time, and your immune system fails to recognize it when you catch your next cold. Usually, the symptoms you experience is your immune system reacting to the virus, and not directly caused by the virus itself.

Cold or flu?

Many of the symptoms of cold and flu are similar, and that makes them hard to differentiate. There are many viruses that cause the common cold, and the influenza virus is caused by three virus subtypes. Generally, the symptoms of a cold are much milder than those of the flu.

Common cold
Frequent sneezing, a mild to moderate fever, low energy, and tiredness are all products of the common cold. There are over 100 viruses responsible for the common cold, and it can be acquired at any time of the year.

Shared symptoms
Both the common cold and flu are classed as upper respiratory tract infections. Either illness may cause a runny nose, sore throat, coughs, headaches, an aching body, shaking, and chills.

Flu
Influenza is caused by virus types A, B, and C. Having the flu may induce a moderate to high fever and constant tiredness. It is generally caught in the winter months and can develop into more serious conditions such as pneumonia.

How a virus invades a cell

Viruses need to invade healthy cells to replicate. A virus tricks the cell into making copies of it. A cell's nucleus is where instructions to make body proteins are stored. Viruses are surrounded by a coat of protein, and the virus can hijack cells to make these viral proteins instead of normal body proteins. Once they have replicated, the virus will then enter other cells in your body and the cycle continues. This process is the same for both the common cold and the flu.

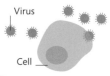

Virus
Cell

1 The virus attaches itself to your cell and the cell engulfs the virus.

Virus
Nucleus of cell

2 Substances in the cell begin to strip the virus's outer coat of protein.

Nucleic acid (DNA or RNA)

3 Nucleic acid from the virus is released, ready to be replicated.

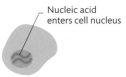

Nucleic acid enters cell nucleus

4 Your cell replicates the viral nuclei acid under the false pretense that it is your own DNA.

Virus has been replicated

5 The cell ignores its own chemical needs and switches to making new viral nucleic acids, which become copies of the virus.

Damaged cell

6 The virus is released from the host cell. This can destroy the cell, and the viruses go on to invade other cells.

A change in your mood can be brought about by the annoyance of having a runny nose and lack of sleep

MOODINESS

HEADACHES
It is thought the chemical cocktail released during an immune response increases pain sensitivity in the brain, causing headaches.

A dilation of blood vessels in the nasal passages, sinuses, and mucus build-up leads to a congested feeling in the head

SINUSES

The inflammation of your sinuses stimulates mucus production in your nasal cavity. The increased mucus forms a barrier against incoming viral cells

RUNNY NOSE

SNEEZING

The release of histamines triggers sneezing, which helps to clear the viral cells out of your nose. However, this also leads to the virus spreading

FEVER

A rise in body temperature is another way that our immune system combats infection. The body's temperature regulation system is reset to a higher level to speed up immune reactions required to fight infection. As long as a fever is mild, there is no cause for worry – but persistant fevers should be monitored.

Immune response

The invasion of viral particles into the epithelial cells found within the mouth or nose triggers an immune response. Symptoms of the common cold or flu are a product of this immune response. The affected epithelial cells release a cocktail of chemicals including histamines, which causes an inflammation of your sinuses, and cytokines, which command cells involved in your immune response.

SORE THROAT

A reflex to clear your airways of mucus build up, coughing may be triggered by inflamed cells and some of the chemicals released as part of the immune response

COUGHING

An inflammation of the epithelial cells in the throat is one of the first symptoms of cold and flu, and so is often understood as a warning sign for when you are "coming down with something"

EXHAUSTION
All of these symptoms will interrupt your sleeping pattern. Cytokines exacerbate the feeling of exhaustion, forcing your body to slow down to fight the virus.

CHILLS
Shivering raises your body temperature – rapid contractions from your muscles generates heat, helping to speed up immune reactions that fight off the infection.

Vaccine action

One of the most effective ways of preventing the spread of infectious disease is to prime the immune system through vaccination. A vaccine trains the immune system to launch a fast and furious attack on a pathogen.

Herd immunity

Vaccinating a significant portion (around 80 per cent) of a population can help provide immunity even to those who have not been vaccinated. When the disease is passed to vaccinated individuals, their primed immune system destroys it, preventing it from spreading further. This can help protect people who can't be vaccinated due to age or illness. Widespread vaccination can eliminate diseases entirely, such as smallpox.

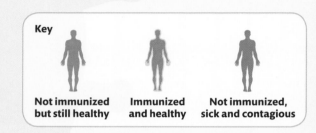

Key

Not immunized but still healthy

Immunized and healthy

Not immunized, sick and contagious

Safety first
Contagious diseases can be contained if a sufficient number of people are vaccinated. Vaccination also helps people who have an existing medical condition that may be worsened by the effects of the disease.

TO VACCINATE OR NOT?

Controversy exists over the use of vaccines. Scares over possible side effects have led some parents to refuse to have their children vaccinated, which has resulted in outbreaks of preventable diseases, such as measles and whooping cough. If only a small portion of the population is vaccinated, herd immunity breaks down.

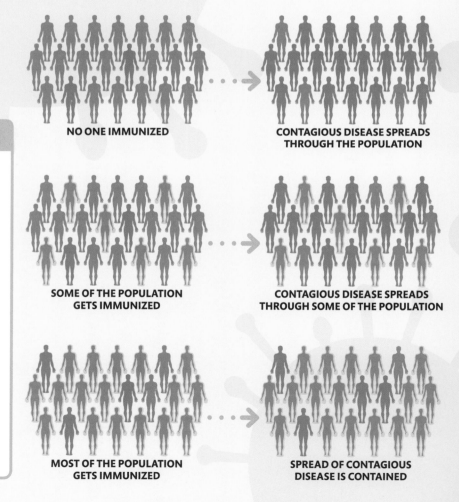

NO ONE IMMUNIZED

CONTAGIOUS DISEASE SPREADS THROUGH THE POPULATION

SOME OF THE POPULATION GETS IMMUNIZED

CONTAGIOUS DISEASE SPREADS THROUGH SOME OF THE POPULATION

MOST OF THE POPULATION GETS IMMUNIZED

SPREAD OF CONTAGIOUS DISEASE IS CONTAINED

Types of vaccines

Each vaccine is developed for a specific pathogen and is designed to kickstart the immune system. This is done by injecting a harmless version of the pathogen that the immune system will remember if attacked by the real pathogen. This can be difficult – killing the pathogen may make it safe, but the vaccine may not produce an immune response. There are also some diseases that progress so quickly, the immune's memory system may not respond in time, so booster immunizations are given to keep reminding the immune system.

Inactivated
The pathogen is killed using heat, radiation, or chemicals. Used for influenza, cholera, and bubonic plague vaccines.

WHY DO VACCINES MAKE YOU FEEL ILL?

Vaccinations stimulate an immune response, which can produce symptoms in some people – but it means the vaccine is doing what it's supposed to.

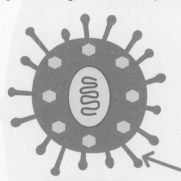

Related microbe
A pathogen that causes disease in another species, but few or no symptoms in humans, is sometimes used. For example, tuberculosis vaccine is made from a bacterium that infects cattle.

ORIGINAL DISEASE-CAUSING PATHOGEN

Alive, but not dangerous
The pathogen is kept alive but the parts that make it harmful are removed or disabled. Used for measles, rubella, and mumps vaccines.

DNA
DNA from the pathogen is injected into the body, whose own cells take up this DNA and start to produce proteins from the pathogen, which triggers an immune response. Used for Japanese encephalitis vaccine.

Tame toxins
Toxic compounds released by the pathogen, which are responsible for the illness, are deactivated using heat, radiation, or chemicals. Used for tetanus and diphtheria vaccines.

Pieces of pathogen
Fragments of the pathogen, such as proteins on the surface of the cell, are used instead of the whole pathogen. Used for vaccines against hepatitis B and human papilloma virus (HPV).

Immune problems

Sometimes the immune system is too reactive – launching attacks on things that aren't harmful and even attacking the body's own cells. Allergies, hayfever, asthma, and eczema are all caused by an oversensitive immune system. Alternatively, the immune system may not be reactive enough, leaving the body vulnerable to infection.

ANAPHYLACTIC SHOCK

Sometimes the immune system launches an extreme panic attack when it encounters an allergen such as a sting or a nut. Symptoms include itchy eyes or face, followed quickly by extreme swelling in the face, hives, and difficulty swallowing and breathing. This is a medical emergency that needs to be treated with an injection of adrenaline, which constricts blood vessels to reduce swelling and relax the muscles around the airways.

ARE FOOD ALLERGIES AN IMMUNE RESPONSE?

Yes. Similar to hayfever, allergies to certain foods cause an inflammatory response from the mouth to the gut. Severe allergies may result in anaphylaxis.

Cartilage erodes

Macrophage

JOINT

Inflamed joint

B cell

Rheumatoid arthritis
If the immune system attacks cells around a joint, causing an inflammatory response, an autoimmune disease called rheumatoid arthritis can result. The joint swells, gets inflamed, and is very painful. Eventually, there is permanent damage to the joints and surrounding tissues.

Immunity overload

Most immune problems are a combination of genetic and environmental factors. While immune conditions are usually triggered by exposure to environmental factors, such as pollen, foods, or irritants on the skin or in the air, some people are genetically more susceptible to developing them. Even autoimmune diseases (when the immune system attacks healthy body tissue by mistake), such as rheumatoid arthritis, can be made worse by irritants that cause inflammation elsewhere in the body. People with a hypersensitive immune system may experience several conditions; for example, many people with asthma also suffer from allergies.

Raised, itchy skin

Hair

Allergen

Epithelium

SKIN

Mast cell releasing histamine

Eczema
The causes of eczema are unclear, but it is thought to be a miscommunication between the immune system and the skin. It is probably triggered by an irritant (allergen) on the skin that stimulates the immune system beneath to launch an inflammatory response, causing swelling and redness.

Allergies and our modern lifestyle
More people in developed countries suffer from allergies, and incidences have been rising since World War II. The specific reasons are subject to debate, but there is agreement that it is likely to do with the immune system being exposed to fewer microbes during childhood.

SINUS

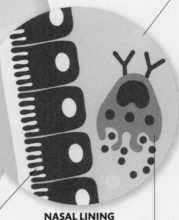

Allergen

Hayfever
Many people have a specific allergy to pollen or dust called hayfever. When allergens bind to the membranes of immune cells just below the epithelium of the eyes and nose, it triggers these cells to release histamines. This triggers an inflammatory response, including itchy, watery eyes and sneezing.

Epithelium

NASAL LINING

Mast cell secretes histamines

Lining of bronchus

Allergen

Cytokines released by immune cell triggers swelling

Immune cell

NORMAL IMMUNE RESPONSE

LUNG

Swollen bronchus

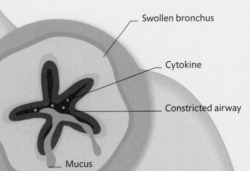

Cytokine

Constricted airway

Mucus

ASTHMA ATTACK

Asthma
An attack of asthma is a spasm in the bronchi of the lungs leading to wheezing, coughing, and breathing difficulties. It is brought on by an allergic response in the lungs to some irritant in the environment. There is some evidence that this condition can be inherited.

WEAKENED IMMUNITY

When the immune system is weakened or absent, a person is said to be immunocompromised. This can happen because of genetic defects, as a result of HIV or AIDS, certain cancers and chronic diseases, and as a consequence of chemotherapy or having to take immunosuppressant drugs after a transplant. People with weakened immunity have to avoid even simple infections, such as colds, because they cannot fight them effectively. Even vaccines pose a risk of causing infection.

BIOHAZARD

CHEMICAL
BALANCE

Chemical regulators

Some of the organs of the endocrine system are dedicated specifically to hormone production, while others, such as the stomach and the heart, have other more familiar functions too. Each receives information from the body and responds by secreting either more or less of a certain hormone. The hormones act as messengers, telling cells to either "keep the balance" or giving instructions to bring about short-term or long-term changes, such as puberty.

SLEEP

Pineal gland
When light levels decrease, the pineal gland releases melatonin, which makes you sleepy. It works in close partnership with the hypothalamus.

NERVOUS SYSTEM

Hypothalamus
The hypothalamus is a part of that brain that links the nervous system to the endocrine system. It sits above the pituitary gland and works with it closely. Among other things, it controls thirst, fatigue, and body temperature.

ENERGY

Thyroid gland
The thyroid secretes hormones that control growth and metabolic rate. It also secretes calcitonin, which encourages calcium storage in the bones.

IMMUNITY

Thymus
The thymus secretes the hormone that stimulates the production of the pathogen-fighting T cells. The gland is most active in babies and adolescents, and shrinks with the onset of adulthood.

Pituitary gland
Despite being the size of a pea, the pituitary is sometimes called the "master gland". It controls the growth and development of tissues as well as the function of several other endocrine glands.

GROWTH

CALCIUM

Parathyroid glands
Four tiny glands attached to the thyroid regulate calcium levels in the blood and bones. They release a hormone that acts on the kidneys, small intestine, and bones to increase blood calcium levels.

HYPOTHALAMUS

PINEAL GLAND

PITUITARY GLAND

THYROID

PARATHYROIDS

THYMUS

Hormone factories

Molecules known as hormones travel throughout the body, triggering changes in tissues that regulate everything from sleep and reproduction to digestion, growth, and pregnancy. They are secreted into the bloodstream by organs that are collectively known as the endocrine system.

Testes
The testes secrete the male hormone testosterone. This plays a role in the physical development of boys, and maintains libido, muscle strength, and bone density in men.

MASCULINITY

TESTES

Stomach
When the stomach is full, cells in its lining secrete gastrin, a hormone that stimulates neighbouring cells to secrete gastric acid. This acid is needed to break food down see pp.142–43).

STOMACH

HEART

KIDNEY

Heart
Tissues in the heart secrete hormones that encourage the kidneys to expel water. This reduces blood volume, and so decreases blood pressure.

ADRENAL GLAND

PANCREAS

KIDNEY

Kidneys
When the kidneys detect low oxygen levels in the blood, they secrete a hormone that stimulates the production of red blood cells in the bone marrow.

Adrenal glands
These produce hormones that govern the "fight or flight" response, such as adrenaline. They also help regulate blood pressure and metabolism, and secrete a small amount of testosterone and oestrogen.

ACTION

DIGESTION

Pancreas
As well as producing digestive enzymes, the pancreas makes insulin and glucagon – hormones that control blood glucose levels (see pp.158–59).

OVARY

Ovaries
The ovaries produce two hormones that govern female reproductive health – oestrogen and progesterone. These regulate the menstrual cycle, pregnancy, and birth.

FEMININITY

How hormones work

Hormones are molecules that act as messengers between the body's organs and tissues. They are released into the bloodstream, so they travel throughout the body, but they can only affect cells that have receptors to pick them up – and each hormone has its own particular receptor. Some receptors float in the cytoplasm of target cells, others line the cell membrane.

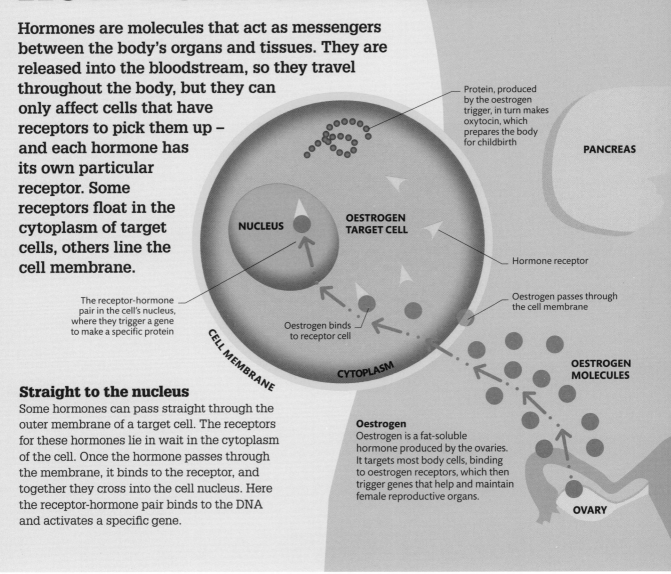

Protein, produced by the oestrogen trigger, in turn makes oxytocin, which prepares the body for childbirth

PANCREAS

NUCLEUS

OESTROGEN TARGET CELL

Hormone receptor

The receptor-hormone pair in the cell's nucleus, where they trigger a gene to make a specific protein

Oestrogen passes through the cell membrane

Oestrogen binds to receptor cell

CELL MEMBRANE

CYTOPLASM

OESTROGEN MOLECULES

Straight to the nucleus

Some hormones can pass straight through the outer membrane of a target cell. The receptors for these hormones lie in wait in the cytoplasm of the cell. Once the hormone passes through the membrane, it binds to the receptor, and together they cross into the cell nucleus. Here the receptor-hormone pair binds to the DNA and activates a specific gene.

Oestrogen
Oestrogen is a fat-soluble hormone produced by the ovaries. It targets most body cells, binding to oestrogen receptors, which then trigger genes that help and maintain female reproductive organs.

OVARY

Hormone triggers

Endocrine glands secrete hormones in response to some sort of trigger. These triggers can be of three kinds; changes in the blood, neural signals, or instructions from other hormones. However, these triggers themselves are often responses to messages from the outside world. When it gets dark, for example, the hormone melatonin is released to help us go to sleep (see pp.198–99).

Triggered by blood
Some hormones are released when sensory cells detect changes in the blood or other body fluids. The parathyroids, for example, release the hormone PTH in response to low calcium levels in the blood (see pp.194–95).

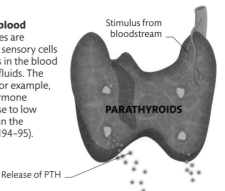

Stimulus from bloodstream

PARATHYROIDS

Release of PTH

TARGET CELLS
CAN HAVE BETWEEN
5,000 AND 100,000
HORMONE RECEPTORS

WHAT IS HORMONE THERAPY?

Hormones can be used to trigger changes throughout the body. Sex hormones, for example, can be manipulated to change individuals to the gender they identify with.

CELL MEMBRANE

Hormone receptor

NUCLEUS

CYTOPLASM

LIVER CELL

GLUCAGON MOLECULES

Glucagon binds to receptor on cell surface

Receptor triggered

A second messenger protein is made due to the glucagon trigger. Its job is to stimulate the liver to make glucose

Glucagon
Glucagon, released by the pancreas, targets liver cells, where it binds to receptors on the cell surface. This prompts the cell's molecular machinery to start converting glycogen into glucose (see pp.156–57).

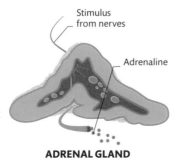

Messenger at the gate

Another class of hormones can't pass through the outer membrane of a cell. These hormones bind to receptors on the surface of the cell instead. This triggers the cell to produce a "second messenger" protein, which causes further changes within the cell.

Triggered by nerves
Many endocrine glands are stimulated by nerve impulses. When we experience physical stress, for example, an impulse is sent along nerves to the adrenal gland, causing it to secrete the fight-or-flight hormone adrenaline (see pp.240–41).

Stimulus from nerves

Adrenaline

ADRENAL GLAND

Triggered by hormones
Hormones can also be released in response to other hormones. The hypothalamus, for example, produces a hormone that travels down to the pituitary gland and prompts it to release a second hormone – growth hormone – which in turn stimulates growth and metabolism.

HYPOTHALAMUS

Hormone stimulus

PITUITARY GLAND

Growth hormone

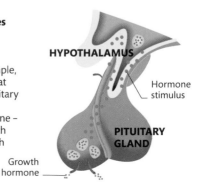

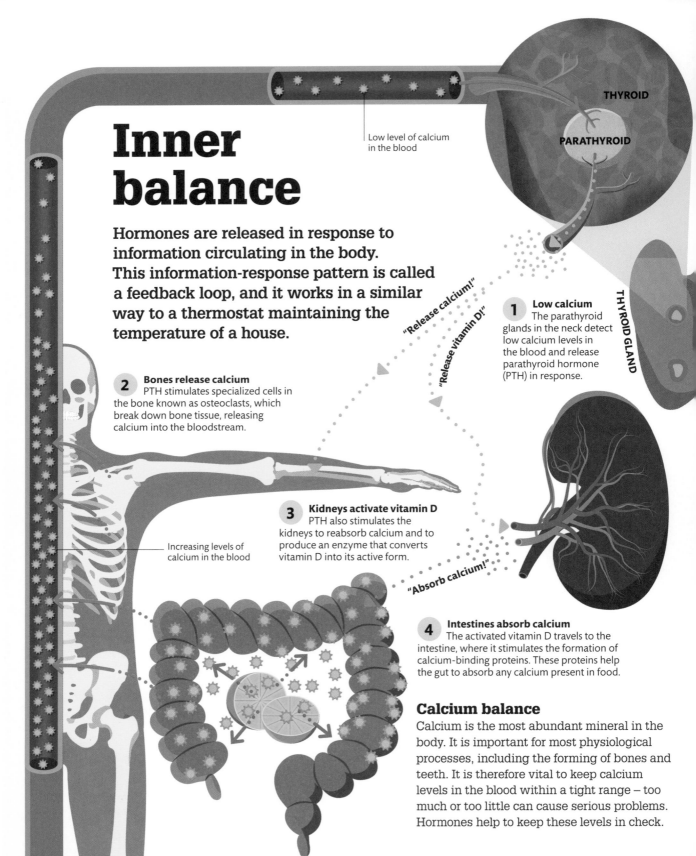

Inner balance

Hormones are released in response to information circulating in the body. This information-response pattern is called a feedback loop, and it works in a similar way to a thermostat maintaining the temperature of a house.

Low level of calcium in the blood

THYROID

PARATHYROID

THYROID GLAND

"Release calcium!"

"Release vitamin D!"

1 Low calcium
The parathyroid glands in the neck detect low calcium levels in the blood and release parathyroid hormone (PTH) in response.

2 Bones release calcium
PTH stimulates specialized cells in the bone known as osteoclasts, which break down bone tissue, releasing calcium into the bloodstream.

Increasing levels of calcium in the blood

3 Kidneys activate vitamin D
PTH also stimulates the kidneys to reabsorb calcium and to produce an enzyme that converts vitamin D into its active form.

"Absorb calcium!"

4 Intestines absorb calcium
The activated vitamin D travels to the intestine, where it stimulates the formation of calcium-binding proteins. These proteins help the gut to absorb any calcium present in food.

Calcium balance

Calcium is the most abundant mineral in the body. It is important for most physiological processes, including the forming of bones and teeth. It is therefore vital to keep calcium levels in the blood within a tight range – too much or too little can cause serious problems. Hormones help to keep these levels in check.

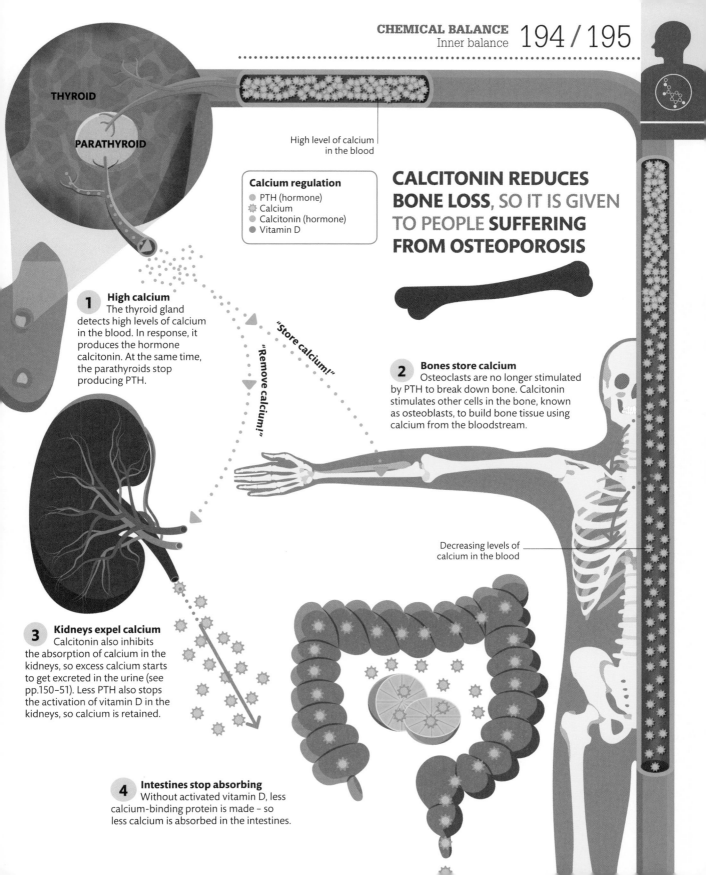

THYROID

PARATHYROID

High level of calcium
in the blood

Calcium regulation
- PTH (hormone)
- Calcium
- Calcitonin (hormone)
- Vitamin D

CALCITONIN REDUCES BONE LOSS, SO IT IS GIVEN TO PEOPLE SUFFERING FROM OSTEOPOROSIS

1 **High calcium**
The thyroid gland detects high levels of calcium in the blood. In response, it produces the hormone calcitonin. At the same time, the parathyroids stop producing PTH.

"Store calcium!"

"Remove calcium!"

2 **Bones store calcium**
Osteoclasts are no longer stimulated by PTH to break down bone. Calcitonin stimulates other cells in the bone, known as osteoblasts, to build bone tissue using calcium from the bloodstream.

Decreasing levels of
calcium in the blood

3 **Kidneys expel calcium**
Calcitonin also inhibits the absorption of calcium in the kidneys, so excess calcium starts to get excreted in the urine (see pp.150–51). Less PTH also stops the activation of vitamin D in the kidneys, so calcium is retained.

4 **Intestines stop absorbing**
Without activated vitamin D, less calcium-binding protein is made – so less calcium is absorbed in the intestines.

Hormonal changes

Hormones often get blamed for our behaviour when the body is undergoing significant change – the moods of a teenager, for example. However, our daily behaviour can also affect our hormones, and that in turn can have serious health implications.

Pituitary gland releases cortisol

Anxiety
People with sedentary lives are less capable of dealing with stress. This may be because they don't have a physical outlet for cortisol and other "fight or flight" hormones that are produced in response to the stresses of modern life.

Smoking affects the function of all the endocrine glands

Sleeplessness and fatigue
Exposure to bright displays such as TVs and phones late at night suppresses melatonin production. This can affect sleep quality and the body's ability to control temperature, blood pressure, and glucose levels.

Pancreas releases copious amounts of insulin

Suppressed immunity
Poor diet and lack of exercise can lead to high cortisol. This hormone is useful in reducing inflammation, but over prolonged periods it can suppress the immune system, which decreases the body's ability to fight infection.

Skin

Unhealthy amounts of fat under the skin

Untoned muscle

High insulin levels
A sedentary life leads to elevated insulin levels, which keeps the body storing fat rather than burning it.

Unhealthy choices

Poor food choices and a sedentary life cause hormone changes that perpetuate that same unhealthy lifestyle. Lower activity levels lead to fewer "feel good" hormones. This can lead to poor food choices, which affect hormones that regulate blood sugar, leading to weight gain and less exercise.

HUGGING RELEASES THE HORMONE **OXYTOCIN**. THIS REDUCES BLOOD PRESSURE SO THE RISK OF **HEART DISEASE FALLS**

Healthy lifestyle

Regular exercise is one of the most effective ways to trigger changes in hormones that lead to a healthier mind and body. Some of the hormones that help equip us for physical activity by regulating temperature, maintaining water balance, and adapting to increased oxygen demands are also so-called "feel good" hormones, which greatly improve mood.

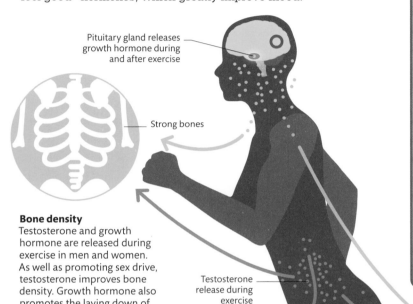

Pituitary gland releases growth hormone during and after exercise

Strong bones

Testosterone release during exercise

Skin

Minimal fat

Lean muscle

Bone density
Testosterone and growth hormone are released during exercise in men and women. As well as promoting sex drive, testosterone improves bone density. Growth hormone also promotes the laying down of bone and continues its work in the night after exercise, encouraging the body to recover and promoting general maintenance.

Good musculature, thanks to growth hormone and testosterone

Muscle mass
Testosterone stimulates the building of lean muscle mass, and increases our overall metabolism. Growth hormone promotes the growth of muscle tissue and helps the body burn fat.

Healthy insulin levels
Insulin is inhibited during exercise, forcing our cells to burn fat as an energy source instead of glucose. Insulin levels remain suppressed for a long time after exercise, meaning we burn fat even as we rest.

Hormones and health
Three hormones play a role in improving our health and our state of our mind.

- - → Growth hormone
- - → Insulin
- - → Testosterone

EXERCISE BUZZ

Exercise increases the release of neurotransmitters, which are the chemical messengers of the nervous system. They transmit signals at junctions between nerve cells, called synapses. The increase promotes the repair and maintenance of the brain. Some neurotransmitters, such as dopamine, also provide a feeling of happiness.

Transmitting nerve cell

Neurotransmitter molecules released

Receiving nerve cell

SYNAPSE BETWEEN TWO NERVE CELLS

Daily rhythms

The body has a built-in time-keeping system that drives our daily rhythms – particularly those of eating and sleeping. At the core of this is the daily chemical conversion of the wakeful hormone serotonin into the sleep hormone melatonin – a process that takes about 24 hours.

The daily cycle

Many hormones go through rhythmic fluctuations every day. These oscillations, happen independently of any external prompting. Even in a black room with no windows, the body gets a serotonin surge in the morning, which wakes it up. However, these rhythms are not hard-wired – they are constantly readjusted and can be changed radically when we travel to a different time zone.

The circadian clock

Our bodies run on a (roughly) 24-hour hormone cycle, known as a circadian rhythm. The biological processes that govern it are called the circadian clock, which is what governs all the body's rhythms. One of the main cogs in this clock is a very small region of the brain known as the suprachiasmatic nucleus (SCN). Located very near the optic nerves, the SCN uses the amount of light entering the eye to calibrate the circadian clock.

Internal timepiece
The SCN drives a two-way chemical conversion between the hormone serotonin, which wakes us up, and melatonin, which puts us to sleep.

3 Hunger hormones
Hunger hormones rise and fall throughout the day. Ghrelin, the appetite increaser, levels rise during fasting, increasing hunger in the morning. Leptin, the appetite suppressor, signals when you are "full".

2 Stress-managing cortisol
As you start the day, the body produces the steroid hormone cortisol, which helps the body deal with stress by increasing blood sugar levels and kick-starting metabolism.

1 Wakeful serotonin
Light stimulates the suprachiasmatic nucleus to convert melatonin into serotonin – a hormone that helps get the brain and body going (especially the intestines).

9AM

8AM

6AM

3AM

The SCN orders the secretion of melatonin or serotonin, depending on the time of day

Light rays of varying intensity

Serotonin

Melatonin

WAKE!

SLEEP!

Electrical signals target the SCN

10 Testosterone surge
Men experience a rise in testosterone levels at night, regardless of whether or not they are asleep – a fact that might explain late-night fights at clubs.

CAN STRESS MAKE YOU ILL?

Stress hormones prepare us for fight or flight, but they also take a toll on some of our other systems, particularly our immune system. Chronic stress can therefore lead to disease.

4 Cortisol peaks
After the morning surge of cortisol, the body gets another dose around noon. From then on, cortisol plays a smaller role in the system. Melatonin is at its lowest level at this time.

Cortisol

Melatonin

12AM

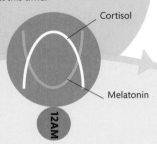

5 Aldosterone surge
Mid-afternoon sees a peak in the hormone aldosterone. This helps to keep the blood pressure steady by increasing water reabsorption in the kidneys.

3PM

JET LAG

Air travel transports us into new time zones faster than the body can adjust. It takes time for the new rhythm of daylight to reset the body clock. Some hormone cycles are more flexible than others – cortisol can take 5–10 days to adapt. While our rhythms adjust, the body feels hungry and sleepy at all the wrong times – a phenomenon called jet lag. Shift workers experience this regularly, and the long-term health consequences are not yet fully understood.

6 Sleepy melatonin
Decreasing light levels prompt the conversion of serotonin into melatonin. This slowly prepares the body for sleep and finally causes sleepiness itself.

6PM

Thyroid gland

8PM

7 Stimulating thyroid
In the evening, levels of thyroid-stimulating hormone abruptly increase. This stimulates growth and repair, but also inhibits neuronal activity, possibly preparing the body for sleep.

9PM

12PM

Melatonin
Cortisol

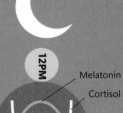

9 Melatonin peaks
Melatonin levels in the blood are highest around midnight. This is also when cortisol levels are at their lowest. This ensures that the body rests completely overnight.

8 Growth hormone
The first two hours of sleep see a burst in growth hormone, which helps children grow and adults regenerate. It's also released in the day, but more is produced at night, when the body can focus on repair.

A BRISK WALK AT
LUNCHTIME HELPS BOOST
SEROTONIN LEVELS

Diabetes

Insulin is the key that opens muscle and fat cells to receive glucose, which the body needs for energy. Without insulin, glucose remains in the blood, and the cells don't get the energy they need, which has serious health consequences. If insulin fails to work, the result is diabetes, a disease that has two forms – type 1 and type 2 – and currently affects 382 million people globally.

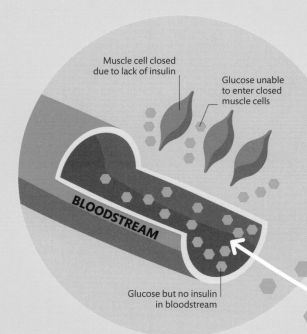

Muscle cell closed due to lack of insulin

Glucose unable to enter closed muscle cells

BLOODSTREAM

Glucose but no insulin in bloodstream

1 **Glucose on the rise**
During digestion, glucose is released into the bloodstream. The rise in glucose levels triggers mechanisms that will lower them – including the release of insulin from the pancreas (see pp.158–59).

Glucose molecule

3 **No entry for glucose**
Without insulin, glucose can't enter the body's cells. Instead, it builds up in the blood, and the body reacts by trying to get rid of it by other means, such as urination.

Type 1 diabetes

In type 1 diabetes, the body's immune system attacks the insulin-producing cells of the pancreas, leaving the pancreas unable to produce any insulin. The symptoms emerge over a matter of weeks but can be reversed once treated with insulin. Although people can develop type 1 diabetes at any age, most are diagnosed before the age of 40, particularly in childhood. Type 1 accounts for 10 per cent of all diabetes cases.

2 **No insulin available**
However, in type 1 diabetes, the insulin-producing cells of the pancreas have been destroyed by the body's own immune cells. As a consequence, no insulin is released to counter the rising glucose levels.

PANCREAS

The symptoms of diabetes

The symptoms of type 1 and type 2 diabetes are similar. The glucose that the kidneys can't get rid of starts to build up in the body, so the body tries to flush it out, so thirst, water intake, and urination increase. Meanwhile, the body's cells are being starved of glucose, which causes fatigue throughout the body. Weight loss also occurs, due to the body burning fat instead of glucose.

Always feeling thirsty, hungry, and tired

Blurred vision caused by build-up of glucose in the lenses

Bad breath caused by ketones being burned instead of glucose (see p.159)

Hyperventilation caused by lack of energy

Weight loss

Nausea and vomitting

Frequent urination

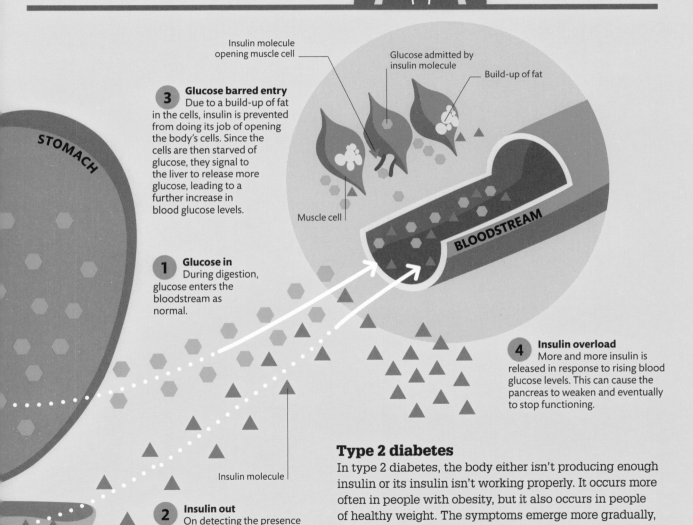

Insulin molecule opening muscle cell

Glucose admitted by insulin molecule

Build-up of fat

3 **Glucose barred entry**
Due to a build-up of fat in the cells, insulin is prevented from doing its job of opening the body's cells. Since the cells are then starved of glucose, they signal to the liver to release more glucose, leading to a further increase in blood glucose levels.

Muscle cell

STOMACH

BLOODSTREAM

1 **Glucose in**
During digestion, glucose enters the bloodstream as normal.

4 **Insulin overload**
More and more insulin is released in response to rising blood glucose levels. This can cause the pancreas to weaken and eventually to stop functioning.

Insulin molecule

2 **Insulin out**
On detecting the presence of glucose in the bloodstream, the pancreas releases insulin.

Type 2 diabetes

In type 2 diabetes, the body either isn't producing enough insulin or its insulin isn't working properly. It occurs more often in people with obesity, but it also occurs in people of healthy weight. The symptoms emerge more gradually, although some people may not show symptoms at all. In fact, 175 million people globally are thought to be living with undiagnosed type 2 diabetes. Type 2 accounts for 90 per cent of all diabetes cases.

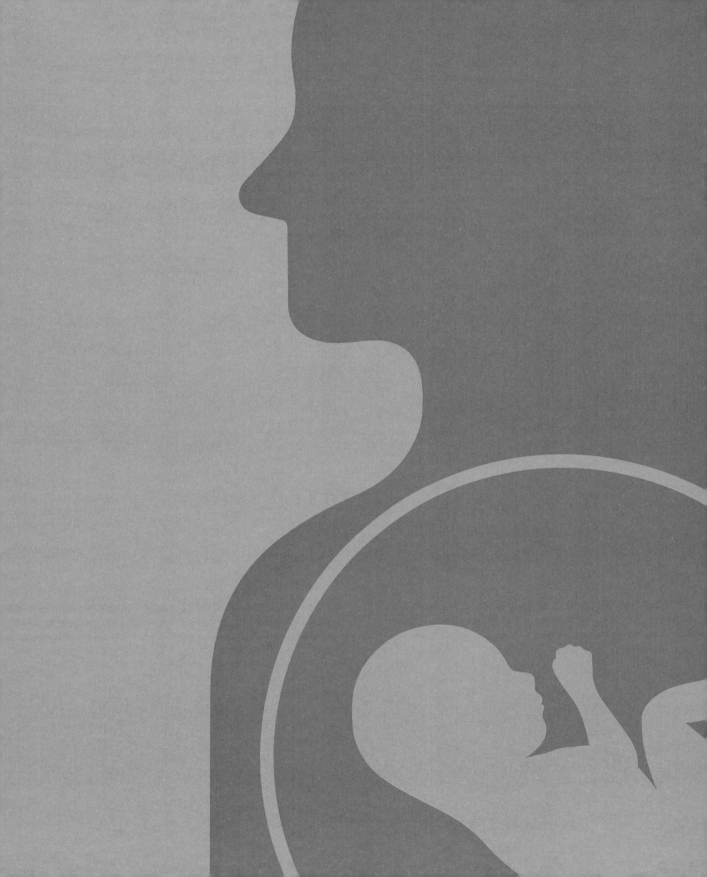

THE CIRCLE

OF LIFE

Sexual reproduction

You are driven by your genes to reproduce, so that your genes continue to multiply in generations to come. Evolutionarily speaking, this is why we have sex. Millions of sperm compete against one another to find one egg and begin the process of creating a new individual.

Bringing sperm and egg together

The main aim of sex is to bring genes from the male and the female together. The male inserts millions of packets of genes in the form of sperm into the female in an attempt to fertilize one of her eggs. If successful, the male's and female's genes mix, generating a new, unique combination of genes in the offspring. To achieve this, both male and female individuals become sexually aroused by one another, which causes some physical changes. Genital organs in both genders enlarge due to increased blood flow, the penis becomes erect, and the vagina secretes a lubricating fluid to aid the penis's entry.

SEMEN NORMALLY **CONTAINS 40–300 MILLION SPERM PER MILLILITRE** (1–8 BILLION PER 0.3 FL OZ)

Seminal vesicle adds fluid to sperm

Prostate gland adds further fluid to sperm to produce semen

Bulbourethral gland neutralizes acidity of urine in urethra to prevent harm to sperm

Sperm travels through penis in the urethra

Sperm matures in epididymis

WHY DO WOMEN HAVE ORGASMS?

Sensitive nerve endings in the clitoris send pleasurable signals to the brain, causing the vagina to contract tightly around the penis, thus ensuring the male ejaculates as much sperm as possible.

HOW DO ERECTIONS WORK?

The penis contains two cylinders of sponge-like tissue, called the corpora cavernosa. When small arteries at the base of the penis dilate, or widen, blood flows into the penis and the corpora cavernosa expand to form rigid cylinders. This compresses small drainage veins so that blood cannot flow away and the penis hardens. After ejaculation, the pressure reduces and drainage veins re-open, allowing blood to flow out and the penis softens.

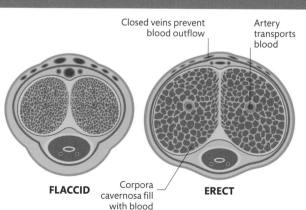

Closed veins prevent blood outflow

Artery transports blood

FLACCID

Corpora cavernosa fill with blood

ERECT

The perilous journey of sperm

During sex, the erect penis is inserted into the vagina. The penis releases semen during orgasm and sperm start their journey to find an egg. Millions of sperm, aided by whip-like movements of their tails, swim up from the vagina, through the cervix, and into the uterus. Sperm is carried in fluid currents caused by the movement of hair-like cells lining the fallopian tubes. Only 150 or so sperm find their way to the upper fallopian tubes, where fertilization usually occurs. The remaining sperm are naturally flushed out of the vagina.

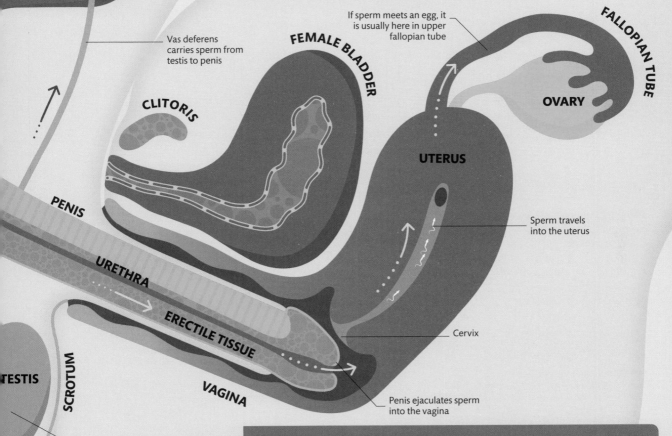

MALE BLADDER

Vas deferens carries sperm from testis to penis

FEMALE BLADDER

If sperm meets an egg, it is usually here in upper fallopian tube

FALLOPIAN TUBE

CLITORIS

OVARY

UTERUS

PENIS

Sperm travels into the uterus

URETHRA

ERECTILE TISSUE

Cervix

TESTIS

SCROTUM

VAGINA

Penis ejaculates sperm into the vagina

Scrotum contains both testes outside body, because sperm production requires a cooler temperature

LARGEST CELL IN THE BODY

An egg (called an ovum) is the largest cell in the human body and just visible to the naked eye. It is protected by a thick, transparent shell. Sperm cells are one of the smallest types of cells in the body, averaging about 0.05 mm (1/500 in) long, but most of this is its tail.

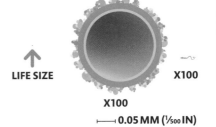

LIFE SIZE

X100

X100

0.05 MM (1/500 IN)

Monthly cycle

Every month, a woman's body prepares for the possibility of pregnancy. Stored in the ovaries, half a million dormant eggs await their turn for ovulation. When hormone levels reach their peak, an egg bursts from an ovary, ready for fertilization. Thick tissue in the uterus lining awaits the egg, if it is fertilized.

Menstrual cycle

The menstrual cycle is controlled by the pituitary gland in the brain. Beginning at puberty, follicle-stimulating hormone (FSH) is produced by the pituitary gland. FSH prompts the production of oestrogen and progesterone hormones in the ovaries. The pituitary gland releases a monthly pulse of FSH and also luteinising hormone (LH), triggering a monthly cycle. A single matured egg is released from the ovary and the lining of the uterus – the endometrium – will thicken and then shed. If the egg is fertilized and then implants into the endometrium, the cycle ceases. Later in life, when the number of dormant eggs in the ovaries reaches a point where they cannot produce enough hormones to regulate the menstrual cycle, menopause is triggered, and the cycle stops.

MENSTRUATION

28 1 2 3 4 5 6 7 8 9 10 11 12 13 14 15 16 17 18 19 20 21 22 23 24 25 26 27

What happens when
The first day of each menstrual bleed is labelled as Day 1. The length of the menstrual cycle varies from woman to woman, but between 21 and 35 days is considered normal. The average length is 28 days.

OVULATION

MENSTRUAL CRAMPS

The muscles in the lining of the uterus naturally contract during a period, constricting tiny arteries in order to limit bleeding. If the contractions are intense or prolonged, they press against nearby nerves, causing pain.

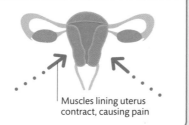

Muscles lining uterus contract, causing pain

3 Hormone surge
Oestrogen is produced by cells in the follicle that surrounds a maturing egg in the ovary. When oestrogen levels peak, this causes a surge of FSH and LH to be released from the pituitary gland, which triggers ovulation.

1 Menstrual bleeding
If a fertilized egg does not implant in the endometrium, falling progesterone levels cut down its blood supply, causing the outer layer to shed as menstrual bleeding. This can serve as an indicator that pregnancy has not occured.

2 Endometrium grows
During the first 2 weeks of the menstrual cycle, steadily rising oestrogen levels cause the endometrium to grow.

OESTROGEN

Bleeding from vagina as endometrium sheds

FSH AND LH

Slight rise in FSH and LH levels stimulates production of oestrogen and progesterone

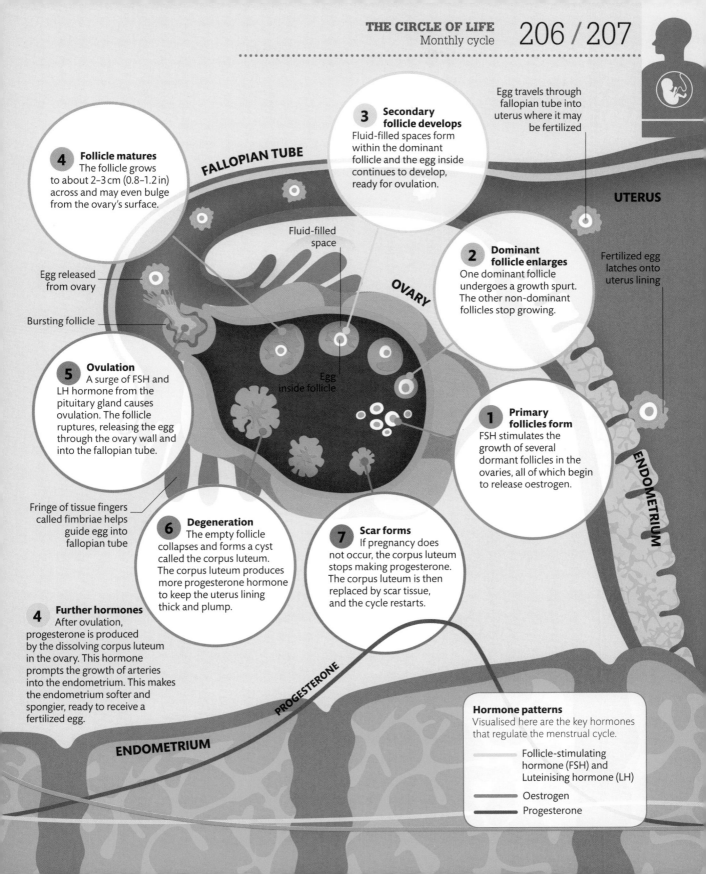

4 Follicle matures
The follicle grows to about 2–3 cm (0.8–1.2 in) across and may even bulge from the ovary's surface.

3 Secondary follicle develops
Fluid-filled spaces form within the dominant follicle and the egg inside continues to develop, ready for ovulation.

Egg travels through fallopian tube into uterus where it may be fertilized

FALLOPIAN TUBE

UTERUS

Fluid-filled space

2 Dominant follicle enlarges
One dominant follicle undergoes a growth spurt. The other non-dominant follicles stop growing.

Fertilized egg latches onto uterus lining

OVARY

Egg released from ovary

Bursting follicle

Egg inside follicle

5 Ovulation
A surge of FSH and LH hormone from the pituitary gland causes ovulation. The follicle ruptures, releasing the egg through the ovary wall and into the fallopian tube.

1 Primary follicles form
FSH stimulates the growth of several dormant follicles in the ovaries, all of which begin to release oestrogen.

ENDOMETRIUM

Fringe of tissue fingers called fimbriae helps guide egg into fallopian tube

6 Degeneration
The empty follicle collapses and forms a cyst called the corpus luteum. The corpus luteum produces more progesterone hormone to keep the uterus lining thick and plump.

7 Scar forms
If pregnancy does not occur, the corpus luteum stops making progesterone. The corpus luteum is then replaced by scar tissue, and the cycle restarts.

4 Further hormones
After ovulation, progesterone is produced by the dissolving corpus luteum in the ovary. This hormone prompts the growth of arteries into the endometrium. This makes the endometrium softer and spongier, ready to receive a fertilized egg.

PROGESTERONE

ENDOMETRIUM

Hormone patterns
Visualised here are the key hormones that regulate the menstrual cycle.

— Follicle-stimulating hormone (FSH) and Luteinising hormone (LH)

— Oestrogen

— Progesterone

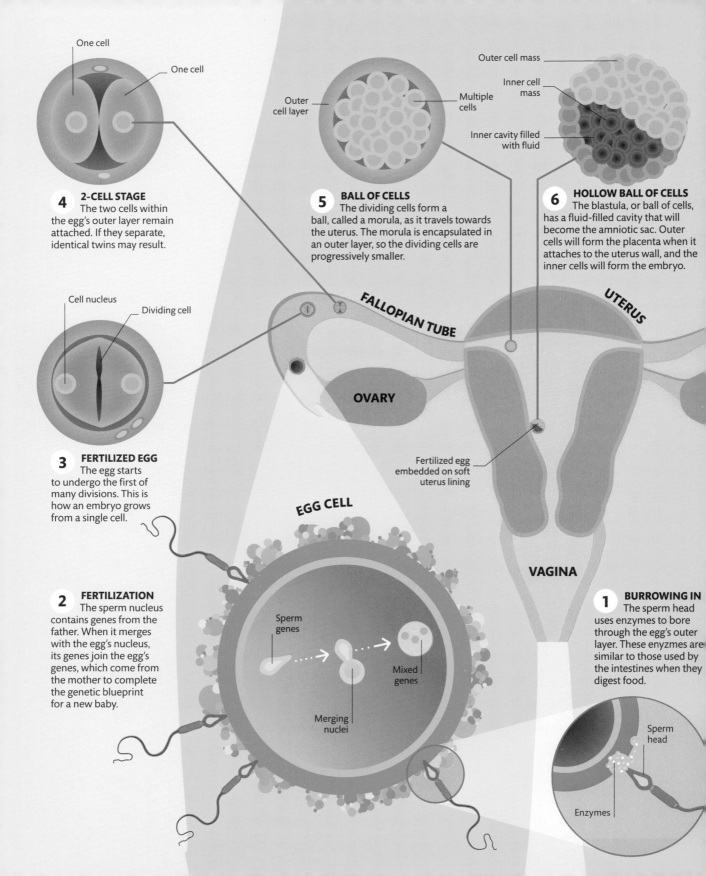

One cell

One cell

4 **2-CELL STAGE**
The two cells within the egg's outer layer remain attached. If they separate, identical twins may result.

Cell nucleus

Dividing cell

3 **FERTILIZED EGG**
The egg starts to undergo the first of many divisions. This is how an embryo grows from a single cell.

Outer cell layer

Multiple cells

5 **BALL OF CELLS**
The dividing cells form a ball, called a morula, as it travels towards the uterus. The morula is encapsulated in an outer layer, so the dividing cells are progressively smaller.

Outer cell mass

Inner cell mass

Inner cavity filled with fluid

6 **HOLLOW BALL OF CELLS**
The blastula, or ball of cells, has a fluid-filled cavity that will become the amniotic sac. Outer cells will form the placenta when it attaches to the uterus wall, and the inner cells will form the embryo.

FALLOPIAN TUBE

UTERUS

OVARY

Fertilized egg embedded on soft uterus lining

EGG CELL

VAGINA

2 **FERTILIZATION**
The sperm nucleus contains genes from the father. When it merges with the egg's nucleus, its genes join the egg's genes, which come from the mother to complete the genetic blueprint for a new baby.

Sperm genes

Mixed genes

Merging nuclei

1 **BURROWING IN**
The sperm head uses enzymes to bore through the egg's outer layer. These enyzmes are similar to those used by the intestines when they digest food.

Sperm head

Enzymes

Tiny beginnings

For some 48 hours after sex, around 300 million sperm race to fertilize an egg as it travels down one of the fallopian tubes. Sperm are chemically attracted to the egg, aiding them on their 15 cm- (6 in-) long journey. When a single sperm fertilizes the egg, a cascade of changes follows.

An egg's journey
Each month, several eggs start to mature within the ovaries. Normally only one developed egg is released at ovulation. The released egg then enters one of the fallopian tubes.

Fertilization

If a woman has ovulated and has had sex, there is a chance of fertilization – the joining of egg and sperm to lay the foundation for pregnancy. The moment the sperm penetrates the egg's outer layer, the egg undergoes a rapid chemical change and hardens to prevent other sperm from burrowing in. Now the combined egg and sperm is called a zygote. It begins to divide as it enters the womb (uterus). Fertilization may have been achieved, but there is a long way to go yet until birth.

WHEN DOES PREGNANCY BEGIN?

Pregnancy does not start until the fertilized egg successfully embeds itself in the soft lining of the uterus – at this point, new life has potentially been conceived.

THE ANSWER TO INFERTILITY

Infertility problems are common in both genders, and affect one in six couples. Some females may have problems with ovulation, their fallopian tubes may be blocked, or their eggs may be too old. Alternatively, males can suffer from a low sperm count, or their sperm may swim poorly. Nevertheless, there are a number of treatments available. One of which, in vitro fertilization, involves collecting eggs and sperm and placing them in a "test tube" for fertilization to occur. The fertilized egg is then allowed to develop before it is implanted back into the uterus to continue its development. A more advanced procedure is intracytoplasmic sperm injection, in which a sperm nucleus is injected directly into an egg.

SPERM   **EGG**

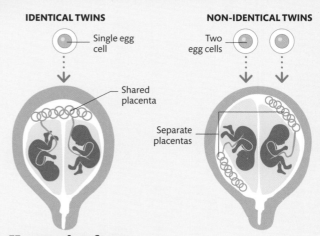

IDENTICAL TWINS

Single egg cell

Shared placenta

NON-IDENTICAL TWINS

Two egg cells

Separate placentas

How twins form

If two eggs are released at ovulation and both are fertilized, non-identical twins result. These can be the same or different genders and each has its own placenta. If a single fertilized egg splits during the early stages of division and each embryo continues to divide separately, then identical twins result, each with their own placenta. If the egg divides late, identical twins share a placenta.

The generation game

Although you are a unique individual, you may have familiar features that are shared by your family. These traits are handed down from generation to generation by genes carried by the mother's eggs and father's sperm.

Hereditary traits

Genes instruct your body on how to develop (see p.23). Structures called chromosomes carry multiple genes (see p.16). Each sperm and egg cell from your father and mother contain a random selection of their genes. When these cells merge during fertilization, the sets of genes mix, forming a new and unique blueprint that makes you "you". If you have brothers or sisters, they inherit a similar selection of genes to you, so you may resemble each other in facial features or body shape and may share similar personality traits or mannerisms. Alternatively, siblings can inherit only a few of the same genes, and may not even look related at first glance.

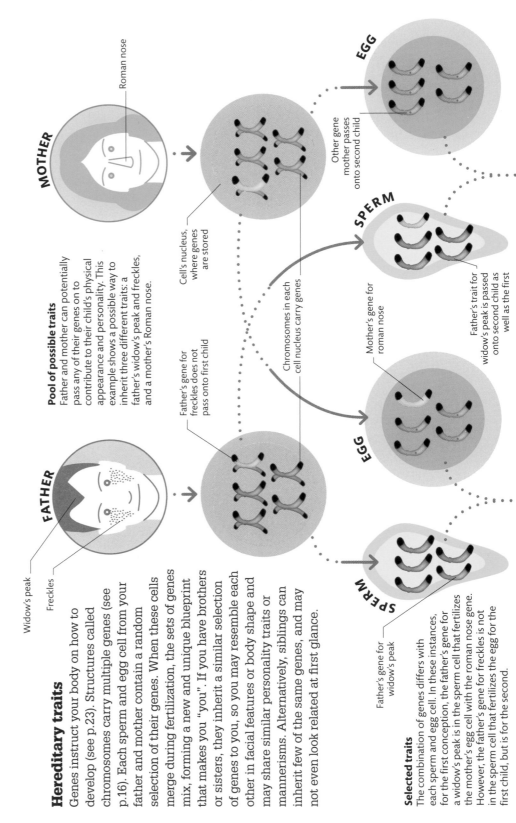

Roman nose

MOTHER

Pool of possible traits
Father and mother can potentially pass any of their genes on to contribute to their child's physical appearance and personality. This example shows a possible way to inherit three different traits: a father's widow's peak and freckles, and a mother's Roman nose.

Cell's nucleus, where genes are stored

Other gene mother passes onto second child

EGG

Father's gene for freckles does not pass onto first child

Chromosomes in each cell nucleus carry genes

Mother's gene for roman nose

SPERM

Father's trait for widow's peak is passed onto second child as well as the first

Widow's peak

Freckles

FATHER

EGG

Father's gene for widow's peak

SPERM

Selected traits
The combination of genes differs with each sperm and egg cell. In these instances, for the first conception, the father's gene for a widow's peak is in the sperm cell that fertilizes the mother's egg cell with the roman nose gene. However, the father's gene for freckles is not in the sperm cell that fertilizes the egg for the first child, but is for the second.

Traits from both parents

The sperm and egg that produced the first child has passed on the father's gene for a widow's peak and the mother's gene for a Roman nose. As a result, this child will share traits with both parents. By chance, they have not inherited their father's freckles.

Shared traits

The second child inherits the father's genes for both a widow's peak and freckles. The siblings share at least one physical characteristic – the widow's peak.

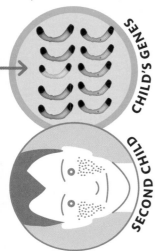

FIRST CHILD

CHILD'S GENES

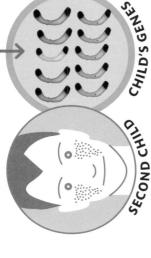

SECOND CHILD

CHILD'S GENES

Dominant and recessive traits

Traits can be inherited in a dominant or recessive pattern. The dominant and recessive versions of a gene are called alleles and are found at the same place on a chromosome. A dominant allele usually shows its trait whenever it is present, while a recessive gene only shows its effects if a more dominant version is absent. If you have detached earlobes, you have at least one dominant allele. Only if you have two copies of the recessive version do you show the recessive trait – the rarer attached earlobe.

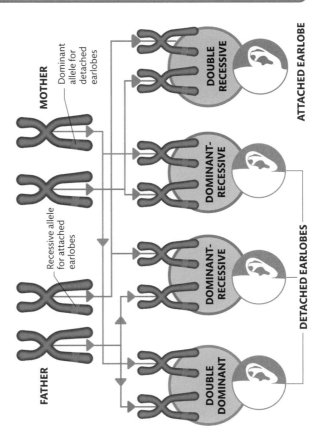

FATHER

MOTHER

Recessive allele for attached earlobes

Dominant allele for detached earlobes

DOUBLE DOMINANT

DOMINANT-RECESSIVE

DOMINANT-RECESSIVE

DOUBLE RECESSIVE

DETACHED EARLOBES

ATTACHED EARLOBE

GENDER-LINKED INHERITANCE

If a mother carries a faulty recessive gene for a vision problem on one X chromosome, her body will use the fully working gene on her other X chromosome. A daughter who inherits the faulty gene will (like her mother) be a carrier and won't be affected, as the dominant gene masks its effects. However, as males have only one X chromosome, any son with the faulty gene will have deficient vision.

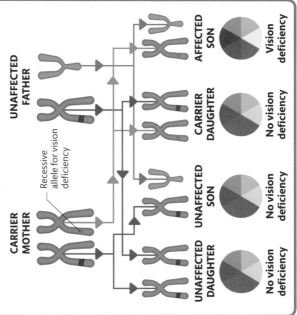

CARRIER MOTHER

UNAFFECTED FATHER

Recessive allele for vision deficiency

UNAFFECTED DAUGHTER

UNAFFECTED SON

CARRIER DAUGHTER

AFFECTED SON

No vision deficiency

No vision deficiency

No vision deficiency

Vision deficiency

Growing life

The development of new life is a miraculous process in which a fertilized egg divides to form a full-grown baby in just 9 months. Connecting the mother and child is the placenta, a special organ that provides the growing fetus with everything it needs.

From cells to organs

During the first 8 weeks, the baby is known as an embryo. Genes switching on and off instruct cells in how to develop. Cells in the outer layer of the embryo form brain, nerve, and skin cells. The inner layer becomes the main organs, such as the intestines, while cells connecting the two layers develop into the muscles, bones, blood vessels, and reproductive organs. Once these main structures are laid down, the baby is called a fetus until birth.

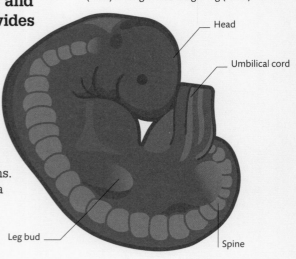

Four-week embryo
The spine, eyes, limbs, and organs have started to form. The embryo is around 5 mm (³⁄₁₆ in) in length and weighs 1 g (¹⁄₃₂ oz).

— Head

— Umbilical cord

Leg bud —

Spine

First heartbeat
Heart growth is almost complete by 6 weeks and all four chambers beat rapidly at around 144 beats per minute. This beating can be detected during an ultrasound scan.

Releases urine
Urine is released by the kidneys into the amniotic fluid every 30 minutes. It is diluted in the fluid and can be swallowed harmlessly by the fetus. Eventually, it passes via the placenta to the mother and she excretes it with her own urine.

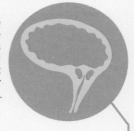

Tiny limbs
The upper limb buds will develop into the arms, while the lower limb buds will form the legs. Fingers and toes begin fused together, then separate.

Lungs form
The two lungs begin to form around this time. They won't be ready to breathe air by themselves until the baby is almost ready to be born.

Fetal development
Every fetus develops at its own rate and the timing for key events tends to vary.

PREGNANCY TIMELINE

1
MONTH

2
MONTHS

3
MONTHS

4
MONTHS

MOTHER

Mother's blood vessel

Mother's blood pools into space

EMBRYO

Embryo's blood vessel

Meeting point
The baby's part of the placenta ends in a fine network of blood vessels that extends into the mother's half of the placenta – close to, but never mixing with, the mother's blood.

Support system
The baby is supported by the placenta – a unique organ that begins growing along with the embryo under the control of both the mother's and the baby's genes. In the placenta, blood vessels from both mother and fetus are intimately interweaved, but the blood never mixes. If it did, the mother's immune system would reject the fetus as "foreign". The fetus gets its oxygen and nutrients from its mother's blood, via the placenta and umbilical cord, in exchange for wastes, such as carbon dioxide.

PLACENTA

AMNIOTIC FLUID

Umbilical cord

Sense of smell
The fetus can recognise the smell of its mother via the amniotic fluid. After birth, the baby is attracted to her smell.

Twitches and "kicking"
A baby's "kick" can be any number of movements the mother feels as her fetus flexes its spine and learns how to move its limbs.

Sensitive to noise
The baby is startled by loud noises. After birth, it will remember songs and voices it heard while inside the womb.

First look
The fetus's eyelids do not open until around the seventh month. When the eyes first open, they cannot form images – they can sense only light and dark.

5	6	7	8	9
MONTHS	**MONTHS**	**MONTHS**	**MONTHS**	**MONTHS**

Mother's new body

The growth of a baby inside a mother's body is an amazing feat – but also a demanding one. The body undergoes an incredible amount of changes and compromises during pregnancy.

Pregnancy transformation

Pregnancy is a time of great physical and emotional change. These changes prepare the mother for the extra demands of pregnancy. The body must not only supply its own needs but also provide the growing baby with all the oxygen, protein, energy, fluid, vitamins, and minerals it needs. The body also absorbs the baby's wastes, and processes them alongside its own. Organs start to support both the body and the baby, so expectant women may tire easily, however the wonder of pregnancy is a remarkable example of the adaptability of the body.

BRAIN

SPINE

LUNG

DIAPHRAGM

Draining brain
The brain recycles its fatty acids in order to provide the baby's brain with the fatty acids it needs. This is a possible cause for the "woolly thinking" many women experience towards the end of the pregnancy. Extra fatty acids in the mother's diet could counteract this problem.

Breasts enlarge
The breasts and nipples enlarge in response to rising levels of the hormone oestrogen. Milk-producing glands in the breast mature in response to progesterone, another hormone. Breasts may start to leak colostrum, or "pre-milk", at the end of pregnancy.

Breathing and heart rates rise
Blood volume increases by about one-third, so the heart pumps harder. Pulse rates of the mother rise, but veins dilate, or widen, so blood pressure naturally falls. Breathing is quicker in order to obtain the extra oxygen the fetus needs.

WHAT CAUSES ODD FOOD CRAVINGS?

Food cravings are undoubtedly one of the strangest phenomena that accompany pregnancy. They may be a symptom of nutritional deficiencies. If the body or the baby is crying out for certain nutrients, this could lead to desires for odd food combinations, such as gherkins with ice cream. Cravings for non-nutritive items such as soil or coal are rarer, but do sometimes occur.

LIVER

STOMACH

Oestrogen

Progesterone

Pressure on spine
As the uterus enlarges, the centre of gravity of expectant women shifts forward so naturally they start to lean back. This alters their posture and puts extra strain on muscles, ligaments, and small joints in the lower spine, causing backache.

Squashed bladder
The bladder is squashed by the rapidly growing uterus, so it holds less urine, resulting in frequent visits to the bathroom. Late in pregnancy, the weight of the uterus stretches the muscles that support the bladder, which can lead to unfortunate leaks when coughing, laughing, or sneezing.

Squashed stomach
As the baby grows, so does the uterus – this pushes the mother's stomach up against the diaphragm. As a result, many expectant women experience heartburn due to acid reflux, and they may be afflicted with loud burps too!

Hormone producer
As it forms, the placenta produces a hormone, human chorionic gonadotropin (hCG), that is detected by pregnancy tests. The placenta then starts to produce oestrogen and progesterone at an increasing rate, causing physical changes such as breast growth.

Abdomen growth
As the uterus expands out of the pelvis, the distance between the pubic bone and the top of the uterus (fundus) helps medics to estimate the stage of pregnancy. A fundal height of 22 cm (9 in) suggests a pregnancy is at around 22 weeks.

WHAT IS MORNING SICKNESS?

Early in pregnancy, hormone changes in the inner ear disrupt the balance of expectant mothers, inducing nausea and causing dizziness similar to when drunk. Morning sickness can happen at any time of day.

THE UTERUS EXPANDS TO **500 TIMES ITS NORMAL SIZE** BY THE END OF PREGNANCY

STRETCH MARKS

Stretch marks are a result of rapid weight gain and stretching of the skin. Deeper in the skin, elastic fibres and collagen that normally keep skin firm and smooth wear thin over the course of pregnancy. Most women are left with stretch marks, however, some lucky women go through pregnancy unscathed.

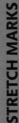

The miracle of birth

Giving birth to a new life is a daunting and exciting experience. Nine months of pregnancy have prepared mother and child for labour – which can last from 30 minutes, up to a few days.

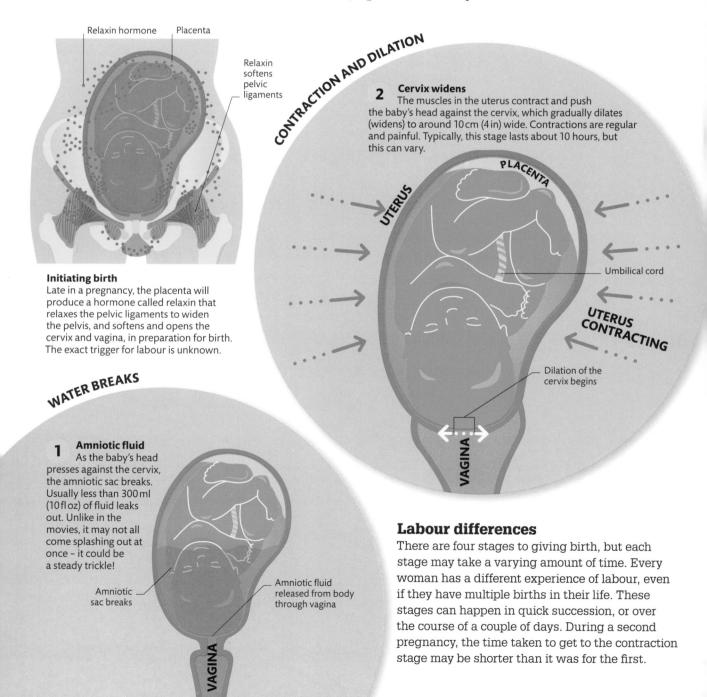

Relaxin hormone Placenta

Relaxin softens pelvic ligaments

Initiating birth
Late in a pregnancy, the placenta will produce a hormone called relaxin that relaxes the pelvic ligaments to widen the pelvis, and softens and opens the cervix and vagina, in preparation for birth. The exact trigger for labour is unknown.

CONTRACTION AND DILATION

2 Cervix widens
The muscles in the uterus contract and push the baby's head against the cervix, which gradually dilates (widens) to around 10 cm (4 in) wide. Contractions are regular and painful. Typically, this stage lasts about 10 hours, but this can vary.

PLACENTA

UTERUS

Umbilical cord

UTERUS CONTRACTING

Dilation of the cervix begins

VAGINA

WATER BREAKS

1 Amniotic fluid
As the baby's head presses against the cervix, the amniotic sac breaks. Usually less than 300 ml (10 fl oz) of fluid leaks out. Unlike in the movies, it may not all come splashing out at once – it could be a steady trickle!

Amniotic sac breaks

Amniotic fluid released from body through vagina

VAGINA

Labour differences

There are four stages to giving birth, but each stage may take a varying amount of time. Every woman has a different experience of labour, even if they have multiple births in their life. These stages can happen in quick succession, or over the course of a couple of days. During a second pregnancy, the time taken to get to the contraction stage may be shorter than it was for the first.

CROWNING

3 Time to push
After a pause, the contractions become more powerful – this is when the mother will feel the need to push. The baby is forced into the vagina (birth canal). Crowning is when the baby's head is first visible.

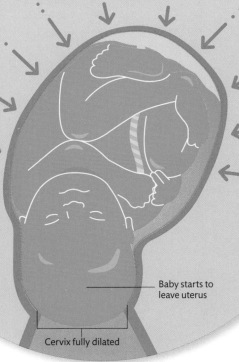

Baby starts to leave uterus

Cervix fully dilated

CARRYING TO FULL-TERM

Pregnancies can vary – only 1 in 20 babies are born on the due date calculated at the beginning of pregnancy. Doctors consider forty weeks as full-term for a single pregnancy, give or take 2 weeks. For twins, doctors consider 37 weeks as full-term, and 34 weeks for a triplet pregnancy. Twins and triplets are born at an earlier stage of their development, and so require extra medical attention.

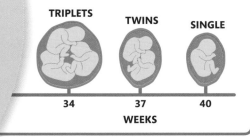

TRIPLETS	TWINS	SINGLE
34	37	40

WEEKS

What happens after birth

After birth, the baby will take its first breath. In doing so, the baby's circulatory and respiratory systems begin to function independently from the mother for the first time. An instant re-routing of blood vessels occurs in order to obtain oxygen from the lungs. The pressure of the blood flowing back to the heart closes a hole in the heart, establishing a normal circulation.

BLOOD CAN BE COLLECTED FROM THE MOTHER'S PLACENTA AND STORED AS A FUTURE SOURCE OF STEM CELLS FOR THE BABY

BIRTH

4 Delivery
Babies are usually born head first. This is so that the widest part of their body, the head, is in line with the widest part of the mother's pelvis, allowing the rest of the baby to pass. The umbilical cord and placenta are delivered during the afterbirth stage.

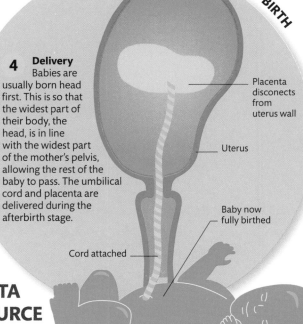

Placenta disconects from uterus wall

Uterus

Baby now fully birthed

Cord attached

Primed for life

You are born with features already in place that will help you grow and develop. Between a newborn's skull bones are flexible, fibrous gaps that allow the head to expand as the brain gets bigger. You will grow fast in your first year and will triple your birthweight.

BABY REFLEXES

Babies are born with over 70 survival reflexes. Placing a finger next to a baby's cheek will make them turn their head and open their mouth. This is the rooting reflex, and it helps them find their mother's nipple when hungry. It fades when regular feeding is established. The grasp reflex helps to stabilize them if they fall, and placing a baby on its stomach will initiate the crawling reflex Both of these are needed for a longer period of time.

ROOTING REFLEX

GRASP REFLEX

CRAWL REFLEX

Months
0 1 2 3 4 5 6 7 8 9

1 MONTH

1 **Starting to smile**
During your first month of life, you listen, watch, and start to recognize people, objects, and places. You will probably smile for the first time at the age of 4–6 weeks.

3 MONTHS

2 **Trying to roll**
At 3 months, you can balance your head, kick and wriggle, and you try to roll over from your back onto your front.

6 MONTHS

3 **Babbling begins**
You speak with babbles and coos. You imitate sounds and respond to simple commands such as "yes" or "no".

9 MONTHS

4 **Sitting up**
You sit up at around 9 months and start to shuffle or crawl. As motor functions develop, you are constantly moving.

Developmental milestones

During the first year of your life, you develop skills that help you explore the world around you. Milestones of development, such as your first smile and first steps, help your carers monitor your progress.

10 MONTHS

5 **Walking tall**
You will start to walk most probably between the ages of 10 and 18 months. Your first steps will occur when holding onto something.

12 MONTHS

6 **Recognizing themselves**
By the age of 12 months, you know your own name, and by 18 months, you start to recognize your own image.

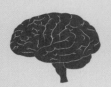

A **NEWBORN'S BRAIN** IS ABOUT **ONE-QUARTER** OF ITS **ADULT SIZE**

Focused senses

A newborn can focus on objects within 25 cm (10 in) and can tell the difference between shapes and patterns. They are familiar with their mother's voice from the womb and they are soothed by gentle, rhythmic noises similar to those of their mother's heartbeat. A baby also recognizes their mother's smell.

3 days
At first, a baby can see only in black and white. A baby finds faces particularly fascinating.

1 month
Normal colour and binocular vision start to develop at the age of about 1 month.

6 months
By 6 months, a baby's vision is excellent. The baby can now distinguish between faces.

Improved dental health when breastfed

Fewer respiratory problems when breastfed

Lower heart rate in breastfed infants

Occurence of food allergies is low when breastfed for 6 months

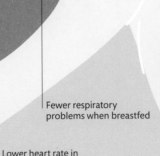

Juvenile arthritis is less common when breastfed

Importance of breastfeeding

Breast milk is the most important source of food for a growing newborn. It is so nutritionally rich that it provides all the energy, protein, fat, vitamins, minerals, and fluids a baby needs during their first 4–6 months. Breast milk also supplies friendly bacteria, conveys antibodies and white blood cells that protect against disease, and delivers essential fatty acids that are vital for development of the brain and eyes. The benefits of breastfeeding are numerous, and influences all of a baby's bones and tissues, and most of their organs.

Understanding others

Between the ages of 1 and 5, most children develop an understanding that other people have their own minds and their own points of view. This is called the "theory of mind". Once a child realises that everyone has their own thoughts and feelings, they can learn to take turns, share toys, understand emotion, and enjoy increasingly complex pretend play as they act out the roles they observe during everyday life.

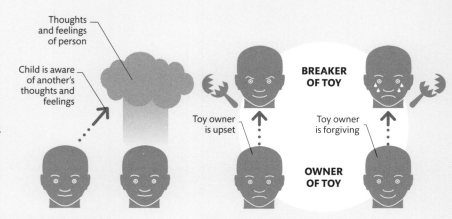

Thoughts and feelings of person

Child is aware of another's thoughts and feelings

BREAKER OF TOY

Toy owner is upset

Toy owner is forgiving

OWNER OF TOY

Understanding others
A child with theory of mind can predict how others might feel in a situation, can understand the intentions behind someone else's actions, and can judge how to respond.

Resentment
Realising a friend broke a toy on purpose causes upset, as the child understands the nefarious intention.

Forgiveness
By recognizing the break was accidental, the child understands that their friend is sorry, and the friendship is secure.

Steady growth

Childhood is a time of rapid physical and emotional growth. Social skills in adulthood are helpful, so it is imperative that children spend time with others of similar ages in order to understand themselves, each other, create boundaries, and establish social bonds. With steady physical growth comes advanced language, emotional awareness, and behavioural rules. New nerve cell connections form in the brain, laying the foundation for mental development.

Childhood development
As you grow, your body proportions develop into a more adult pattern. Growth slows between the ages of 5 and 8.

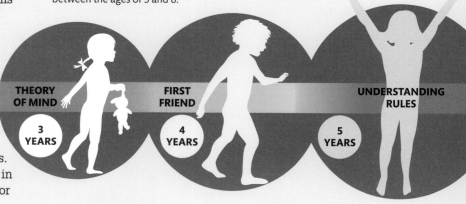

THEORY OF MIND

3 YEARS

FIRST FRIEND

4 YEARS

UNDERSTANDING RULES

5 YEARS

Growing up

You are full of curiosities and energy as a child. During the key stages of childhood until puberty, you gain a good grasp of language, understand that others have minds of their own, you learn about the emotions of others, and you actively start to explore your environment.

CHILDREN AGED BETWEEN 2 AND 10 **ASK** ABOUT 24 QUESTIONS EVERY HOUR

Forming friendships

Many children aged 4 and above now build selective friendships with others who share similar interests and activities. They now have a sense of future, and so can understand the value of a friendship with someone reliable with whom secrets can be shared.

FIRST FRIENDSHIP   **FIRST FALL OUT**   **FIRST MAKE UP**

First resolution
Possessing a theory of mind helps friendships last. When they fall out, children can make up by reflecting on what made their friend upset in order to resolve the conflict.

Understanding rules

Rule-based games help children aged 5 and above to balance their desire to win while following the rules, which discourages cheating and bad behaviour. This helps them recognise right from wrong and how society works later in life.

Rule-abiding behaviour is rewarded

BREAKING THE RULES

FOLLOWING THE RULES

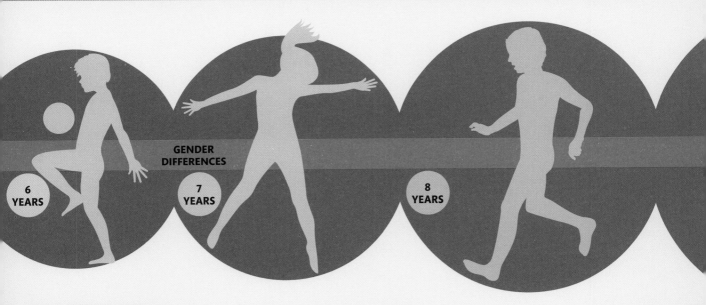

GENDER DIFFERENCES

6 YEARS

7 YEARS

8 YEARS

Friendship groups

Boys and girls have different types of friendship groups by the age of 7, each with their own hierarchy. Boys tend to form large groups of friends comprised of a leader, an inner ring of his close friends, and peripheral followers. On the other hand, girls usually have one or two close friends with equal status. The most popular girls are highly sought after as "best" friends.

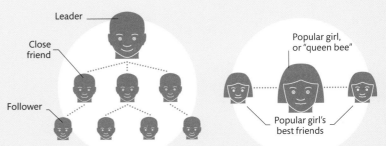

Leader

Close friend

Follower

Popular girl, or "queen bee"

Popular girl's best friends

MALE FRIENDSHIPS

FEMALE FRIENDSHIPS

Hormonal teenagers

Puberty is the stage between childhood and adulthood, when the sex organs mature and reproduction becomes possible. Fluctuating hormone levels cause emotional and physical changes which can make teenagers feel clumsy, moody, and self-conscious.

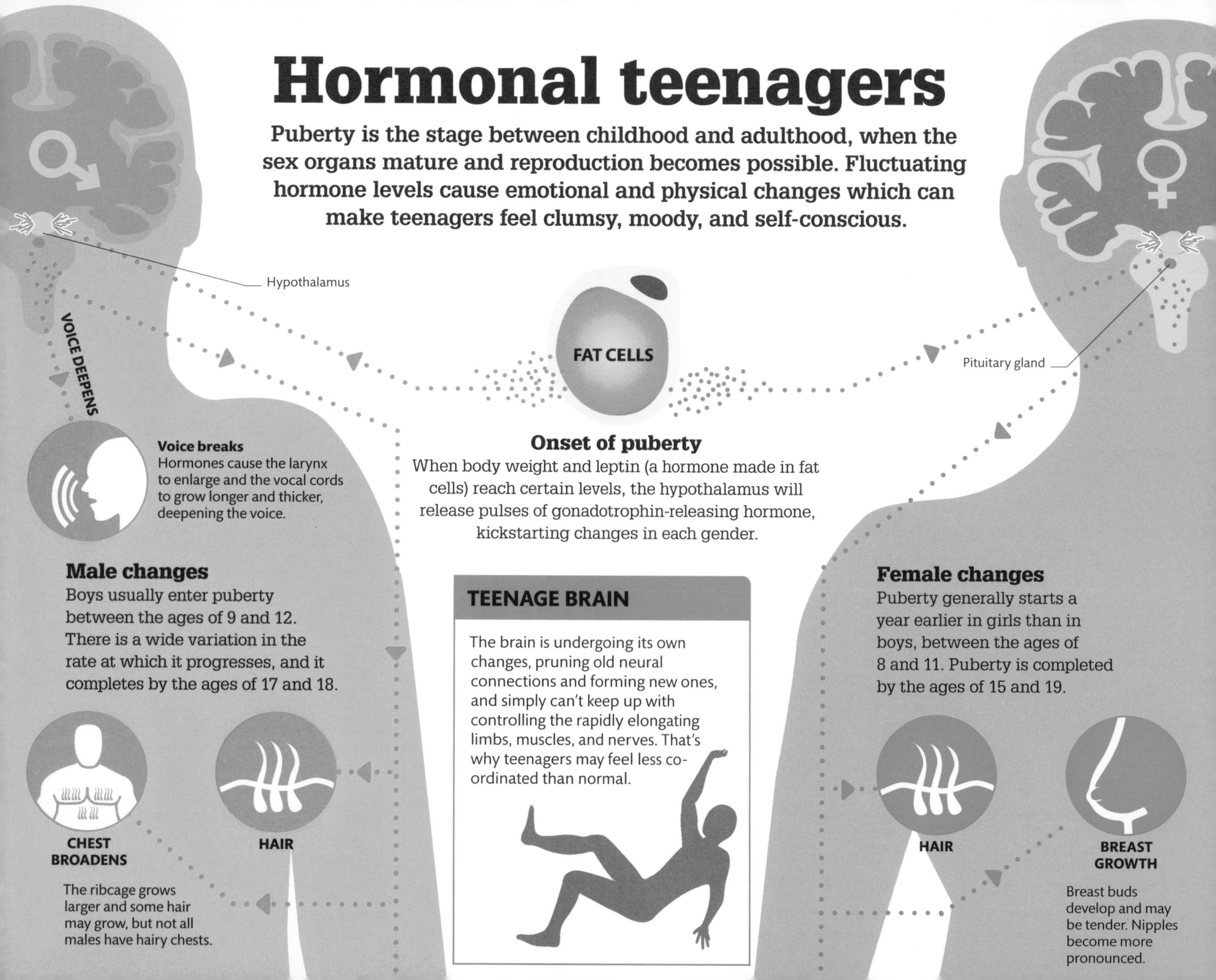

Pituitary gland

Hypothalamus

FAT CELLS

Onset of puberty

When body weight and leptin (a hormone made in fat cells) reach certain levels, the hypothalamus will release pulses of gonadotrophin-releasing hormone, kickstarting changes in each gender.

TEENAGE BRAIN

The brain is undergoing its own changes, pruning old neural connections and forming new ones, and simply can't keep up with controlling the rapidly elongating limbs, muscles, and nerves. That's why teenagers may feel less co-ordinated than normal.

Female changes

Puberty generally starts a year earlier in girls than in boys, between the ages of 8 and 11. Puberty is completed by the ages of 15 and 19.

HAIR

BREAST GROWTH

Breast buds develop and may be tender. Nipples become more pronounced.

Male changes

Boys usually enter puberty between the ages of 9 and 12. There is a wide variation in the rate at which it progresses, and it completes by the ages of 17 and 18.

Voice breaks

Hormones cause the larynx to enlarge and the vocal cords to grow longer and thicker, deepening the voice.

VOICE DEEPENS

HAIR

CHEST BROADENS

The ribcage grows larger and some hair may grow, but not all males have hairy chests.

UTERUS AND OVARIES

PUBIC HAIR

Ovaries produce oestrogen, accelerating puberty changes

Menstruation begins
The first period occurs between the ages of 10 and 16, at an average of 12 years. Ovulation occurs irregularly, and the uterus grows to the size of a clenched fist.

Vaginal secretion
The vagina lengthens and starts to secrete a clear or cream-white discharge – one of the first signs of puberty. The teenager's natural odour may also become stronger.

WHY DO TEENS GET ACNE?

The skin's sebaceous, or oil, glands are stimulated into action by the hormones of puberty. When newly active, they take a while to settle down to a normal rate of oil secretion, so during puberty, many teens suffer from spots.

EARLY AND LATE DEVELOPERS

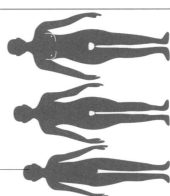

Less developed than her peers of same age

12-YEAR-OLD GIRLS

Puberty starts at different ages, so some friends of the same age may be taller and seem more mature than others. Therefore, three girls at the age of 12 may differ drastically in height and weight. Girls tend to develop earlier than boys because a lower weight of around 47 kg (105 lb) seems to be the key to triggering female puberty. A higher weight average of around 55 kg (120 lb) appears to be the trigger in boys.

DURING A PUBERTY GROWTH SPURT, HEIGHT MAY **INCREASE** BY AS MUCH AS **9 CM** (3½ IN) **IN A YEAR!**

Testes produce testosterone, accelerating puberty changes

PUBIC HAIR

SPERM PRODUCTION IN TESTES

First ejaculation
The penis and testes grow and sperm production begins. The first ejaculation occurs, typically during sleep as a "wet dream".

Getting older

Ageing is a slow and inevitable process. The rate at which you age depends on interactions between your genes, diet, lifestyle, and environment.

Why do you age?

Why ageing occurs is a mystery. We know that the cells in your body divide to renew themselves, but they can only do so a certain number of times. This limit is linked to the number of repeating units, called telomeres, on the end of each of their chromosomes, the X-shaped packages of DNA in every cell's nucleus. If you inherit long telomeres, your cells can undergo more divisions, and as a result you may live longer.

FREE RADICALS

Premature ageing can result from genetic damage caused by free radicals. These molecular fragments are produced by sunlight, smoking, radiation, or pollution damaging your DNA. Dietary antioxidants found in fruit and vegetables help to neutralize free radicals and offer an increased chance of living a longer life.

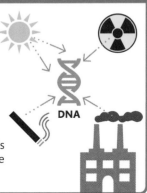

DNA

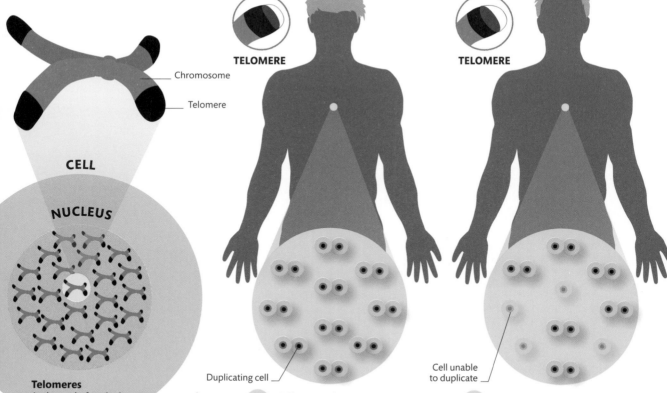

Chromosome

Telomere

CELL

NUCLEUS

TELOMERE

TELOMERE

Duplicating cell

Cell unable to duplicate

Telomeres
At the end of each chromosome arm is a telomere, a repeating section of DNA. During cell division, enzymes attach to the telomeres. These enzymes speed up the chemical reactions involved in cell division.

1 Cell renewal
The enzymes lock on to the telomeres, ready to copy each cell. When an enzyme detaches, it takes a section of telomere, so chromosomes shorten with each division.

2 Depleting telomeres
Eventually, the telomere becomes too short for enzymes to lock into place. The cells with these short telomeres can no longer duplicate or renew themselves. Cells run out of telomeres at different rates.

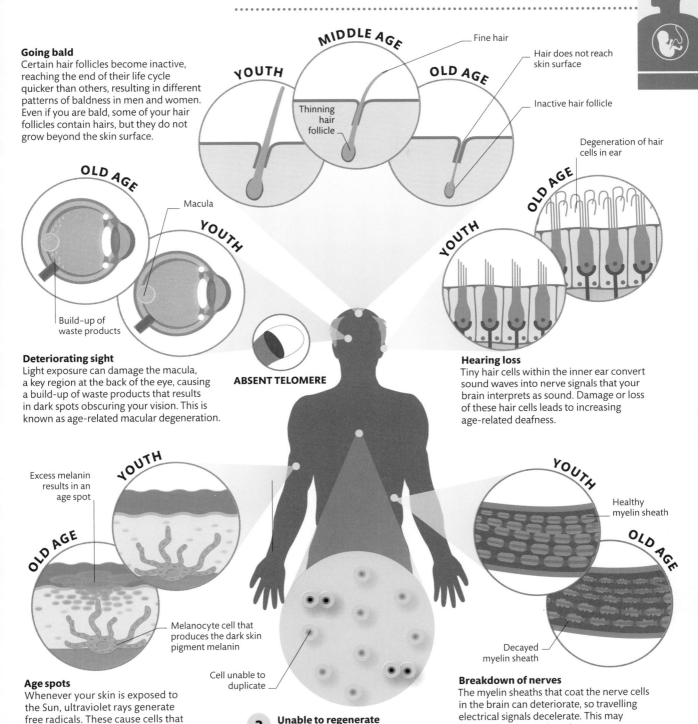

Going bald
Certain hair follicles become inactive, reaching the end of their life cycle quicker than others, resulting in different patterns of baldness in men and women. Even if you are bald, some of your hair follicles contain hairs, but they do not grow beyond the skin surface.

MIDDLE AGE

YOUTH

OLD AGE

Fine hair

Hair does not reach skin surface

Inactive hair follicle

Thinning hair follicle

Degeneration of hair cells in ear

OLD AGE

Macula

YOUTH

YOUTH

Build-up of waste products

Deteriorating sight
Light exposure can damage the macula, a key region at the back of the eye, causing a build-up of waste products that results in dark spots obscuring your vision. This is known as age-related macular degeneration.

ABSENT TELOMERE

Hearing loss
Tiny hair cells within the inner ear convert sound waves into nerve signals that your brain interprets as sound. Damage or loss of these hair cells leads to increasing age-related deafness.

Excess melanin results in an age spot

YOUTH

OLD AGE

Melanocyte cell that produces the dark skin pigment melanin

Age spots
Whenever your skin is exposed to the Sun, ultraviolet rays generate free radicals. These cause cells that produce pigmentation to increase their production, creating age spots.

YOUTH

Healthy myelin sheath

OLD AGE

Decayed myelin sheath

Cell unable to duplicate

3 **Unable to regenerate**
Only a few replicating cells remain in old age. When cells can no longer renew themselves, they slowly deteriorate and the signs of ageing become clear. Cells may die and be replaced with scar tissue or fat.

Breakdown of nerves
The myelin sheaths that coat the nerve cells in the brain can deteriorate, so travelling electrical signals decelerate. This may account for slower thoughts, poor memory recall, and reduced sensation.

The end of life

Death is an inevitable part of the cycle of life. It occurs when all the biological functions that sustain living cells cease. Some deaths result from old age, while some are due to disease and injury.

Leading causes of death
Listed here are the leading causes of death worldwide in 2012, provided by WHO (World Health Organization).

High blood pressure – 4%
Unchecked and uncorrected high blood pressure can be fatal late in life.

Diarrhoeal diseases – 5%
Those suffering from chronic diarrhoea are at risk of fatal dehydration and malnutrition.

What can kill us
Non-infectious diseases, such as heart and lung disease, cancer, and diabetes are most commonly cited on death certificates. Many of these are related to an unhealthy diet, lack of exercise, and smoking, but some are due to nutrient deficiencies.

Lung infections and failures – 16%
Lung cancers and lower respiratory infections together made the second-largest killer in 2012.

HIV – 5%
Deaths caused by the Human Immunodeficiency Virus is decreasing year by year.

Road accidents – 5%
Casualties on the road killed a large number of people in 2012.

Heart and circulation conditions – 60%
Heart attacks and strokes are the two leading causes of death worldwide.

Diabetes - 5%
Those with diabetes may die due to heart disease or stroke because of their condition.

HOW DOES WEALTH AFFECT LIFESPAN?

In high-income countries, 7 in every 10 deaths are among people aged 70 years or older, who've lived a good, long life. In the poorest countries, 1 in 10 children still die in infancy.

1 PER CENT OF THE WORLD'S POPULATION DIES EVERY YEAR

Brain activity

One way to determine if a person is dead is to scan for brain activity. Brain death is diagnosed when electrical recordings (EEG) show an irreversible loss of all higher and lower brain functions, so there is no spontaneous breathing or heartbeat. Someone who is "brainstem dead" can only remain alive if artificial life support is in place.

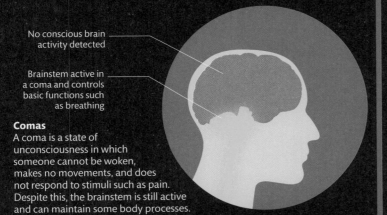

No conscious brain activity detected

Brainstem active in a coma and controls basic functions such as breathing

Comas

A coma is a state of unconsciousness in which someone cannot be woken, makes no movements, and does not respond to stimuli such as pain. Despite this, the brainstem is still active and can maintain some body processes.

Near-death experiences

People who almost die and are then resuscitated often report experiencing similar sensations, such as levitation, looking down on their body, and seeing a bright light at the end of a tunnel. Other common descriptions of such near-death experiences include having flashbacks, or vivid memories, of their earlier life, and being overcome by strong emotions, such as joy and serenity. The cause of these experiences may be changing oxygen levels, sudden release of brain chemicals, or surges of electrical activity – no-one really knows.

YOUR BODY AFTER DEATH

When the heart stops pumping blood, the body's cells no longer receive oxygen or have toxins removed. Chemical changes in muscle cells and the overall cooling of the body cause the limbs of a corpse to stiffen after an initial period of floppiness. This stiffening is called rigor mortis and it wears off again after 2 days.

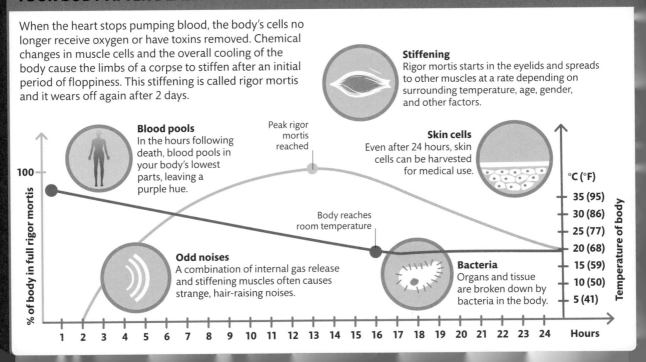

Stiffening
Rigor mortis starts in the eyelids and spreads to other muscles at a rate depending on surrounding temperature, age, gender, and other factors.

Blood pools
In the hours following death, blood pools in your body's lowest parts, leaving a purple hue.

Peak rigor mortis reached

Skin cells
Even after 24 hours, skin cells can be harvested for medical use.

Body reaches room temperature

Odd noises
A combination of internal gas release and stiffening muscles often causes strange, hair-raising noises.

Bacteria
Organs and tissue are broken down by bacteria in the body.

% of body in full rigor mortis

100

°C (°F)
35 (95)
30 (86)
25 (77)
20 (68)
15 (59)
10 (50)
5 (41)

Temperature of body

1 2 3 4 5 6 7 8 9 10 11 12 13 14 15 16 17 18 19 20 21 22 23 24 Hours

MIND
MATTERS

Basis of learning

When we learn a new fact, ability, or react to stimuli, connections between nerve cells form. Messages pass from one cell to another using neurotransmitters (chemicals that are released by nerve cells). The more frequently we remember what we have learnt, the more messages the cells send, and their connection becomes stronger.

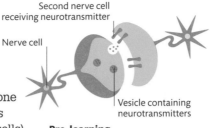

Second nerve cell receiving neurotransmitter

Nerve cell

Vesicle containing neurotransmitters

Pre-learning
Initially, when the nerve cell fires, a small amount of neurotransmitter is released, and there are only a few receptors on the receiving nerve cell.

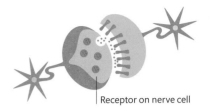

Receptor on nerve cell

After learning
The nerve cell releases more neurotransmitter and more receptors have formed on the second nerve cell, strengthening the connection.

Types of learning

We learn information in different ways, depending on what it is and how it is presented. For some abilities, we have a "critical period" during which we can fully master the skill. For adults who have learnt a new language later in life, they have missed the critical period of acquiring the basic sounds of the language, and therefore may speak with an accent.

LEARNING WHAT TO IGNORE

Unimportant signals
When a stimulus is new, we automatically pay attention to it. If it doesn't signal anything important, we learn to ignore it.

Startled at sound

No response to sound

BEHAVIOUR REINFORCEMENT

Rewards and reprimands
Getting rewarded for good behaviour and reprimanded for bad behaviour can help reinforce our concepts of what is acceptable, and what isn't.

Rewarded behaviour

Behaviour leading to a reprimand

LEARNING BY ASSOCIATION

Associative learning
When two events coincide on a regular basis, we learn to associate them. If you consistently eat when a bell rings, hearing the bell may stir your appetite.

Hunger caused by combined stimuli

Sound alone causes hunger

Learning skills

Connections between nerve cells in your brain allow learning to happen constantly, often with no conscious effort – repetition helps to retain these skills.

EXPLORING A NEW CITY INCREASES YOUR BRAIN SIZE BY FORMING NEW NERVE CELL CONNECTIONS.

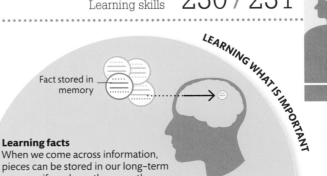

AT WHAT AGE DO WE LEARN THE MOST?

When you are a child, your cognitive, motor, and language skills advance in leaps and bounds – at the age of 2, you tend to learn 10–20 words per week.

LEARNING WHAT IS IMPORTANT

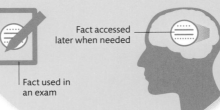

Fact stored in memory

Learning facts
When we come across information, pieces can be stored in our long-term memory if we deem them worth remembering. The judgement can be either conscious or subconscious.

Fact accessed later when needed

Fact used in an exam

LEARNED MOVEMENT (MOTOR SKILLS)

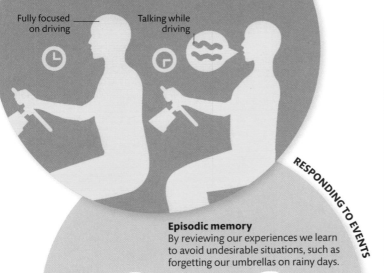

Becoming automatic
When you learn to drive, you concentrate on your movements as well as the traffic. Through repetition, the driving body movements are learned and become automatic, allowing you to give attention to other things at the same time.

Fully focused on driving

Talking while driving

RESPONDING TO EVENTS

Episodic memory
By reviewing our experiences we learn to avoid undesirable situations, such as forgetting our umbrellas on rainy days.

Experience of getting wet

Memory of past experience changes behaviour

EXAM REVISION

When a memory starts to fade, revising the information increases the memory's strength with each revision session – this ensures the learnt information is stored in our long-term memory. Revising information little and often is best for retention. When you cram for an exam or presentation, you acquire a lot of information quickly, but it is lost without revisiting the information – this is why intense study is only useful in the short-term.

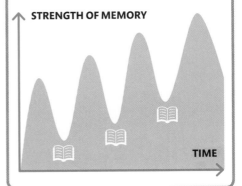

STRENGTH OF MEMORY

TIME

Making memories

Every time you experience something, your brain forms a memory. Inconsequential moments and life-changing events are all stored, but how often you revisit the memory determines whether it is remembered or forgotten. Memories are temporarily stored in your short-term memory, then, if important, transferred to your long-term memory.

WHY DO WE EXPERIENCE DÉJÀ VU?

A sense of familiarity in unfamiliar situations may be because a similar memory is recalled but is confused with the present, so a sense of recognition comes without a concrete memory.

TOUCH

HEARING

SMELL

SIGHT

TASTE

1 Sensory memory
When you sense something, you create a transitory memory, even if you are not conscious of it. It is stored in your sensory memory, and unless transferred to short-term memory, it fades in less than a second.

2 Nerve signals
Encoding is the process in which a sensory memory forms a true memory. When you pay attention to your sensory memory, it enters your consciousness, and the nerve cells that encode the memory fire more rapidly. Nerve cell connections strengthen temporarily to form a short-term memory.

ENCODING

3 Consolidation
New experiences are compared against memories to provide context for new memories. Memories that have emotions and importance attached to them are stronger and less likely to be lost. Sleep is vital for consolidation to happen effectively.

Previous memories provide context

CONSOLIDATION

Final memory

Short-term memory

Our short-term memory can retain around five to seven pieces of information. These memories, such as telephone numbers or directions, are stored only for as long as you need them. Repeating it to yourself helps prolong the memory, but if distracted, you often forget it. Short-term memory is thought to be based on temporary patterns of activity in the brain's prefrontal cortex.

Nerve cell

MEMORY IS FORMED

MEMORY FORGOTTEN

Unimportant memories are lost

Long-term memory

As far as we know, your long-term memory allows you to store unlimited amounts of information. Memories that are most likely to be stored for life include those with a high emotional impact, like a wedding, and that have a semantic value, such as your spouse's name. These memories are connected to growth in areas of the brain linked with memory, such as the hippocampus, so are more stable than short-term memories.

MONTHS

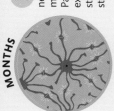

2 Storage
Months later, your nerve cell connections may become permanent. Particularly memorable experiences can jump straight into long-term storage the same day.

YEARS

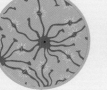

3 Memory fades
If months or years pass before you recall a memory, it is more likely to fade. Specific details about special events, such as the food you ate at your wedding, may be forgotten.

DECADES

4 Losing a memory
Eventually, memories fade – even important ones! It is not known if the nerve cell connections of a memory disappear, or whether they still exist and you are unable to access them.

MEMORY

Nerve cell connection

MEMORY FORGOTTEN

MEMORY RECALL

1 Revisiting a memory
When you recall a memory, the nerve cells encoding it are re-activated. Each time this happens, more nerve cell connections are created and existing ones are strengthened, and the memory is less likely to be forgotten. If you don't recall the memory frequently, it is more likely to be lost.

Nerve cell connection strengthening

STORED MEMORY

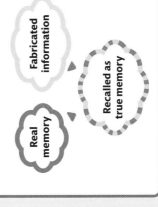

HOLIDAYS

BIRTHDAYS

DATES

JOURNEYS

HOME LIFE

RELATIONSHIPS

MEMORY CONFABULATION

When you recall a memory, the memory enters a labile, or easily altered, state. In a process called confabulation, you may unintentionally add new information to the labile memory when it is reconsolidated. This new information will become an inseparable part of your memory.

Fabricated information

Real memory

Recalled as true memory

Falling asleep

Sleep is a curious phenomenon – we do it every day, but we don't know why. It might allow your body and brain time to repair themselves, flush out toxins that accumulate throughout the day, or strengthen memories. Depriving yourself of sleep is taxing for your body.

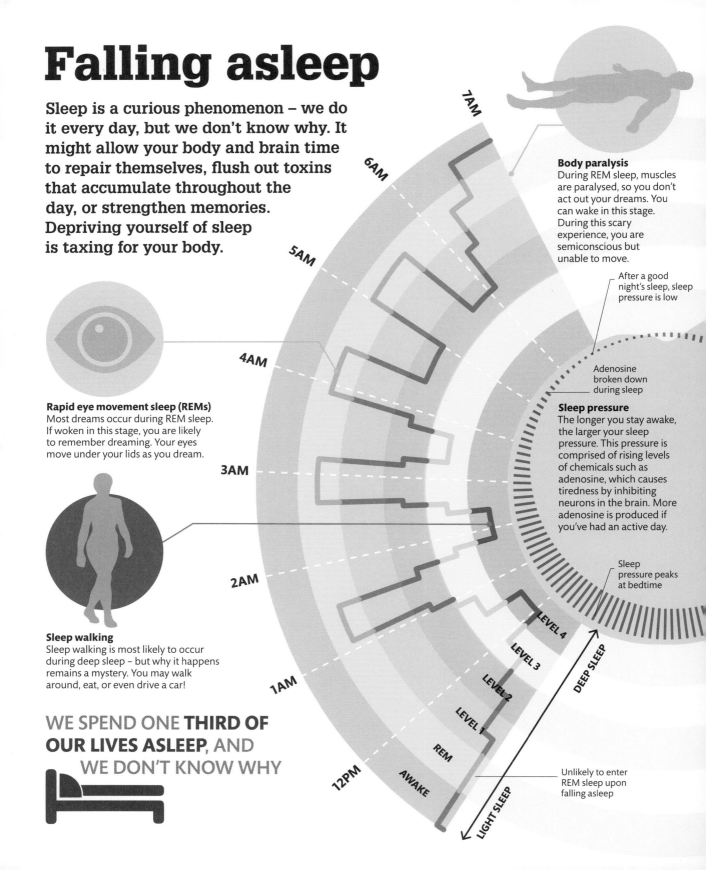

Rapid eye movement sleep (REMs)
Most dreams occur during REM sleep. If woken in this stage, you are likely to remember dreaming. Your eyes move under your lids as you dream.

Sleep walking
Sleep walking is most likely to occur during deep sleep – but why it happens remains a mystery. You may walk around, eat, or even drive a car!

WE SPEND ONE THIRD OF OUR LIVES ASLEEP, AND WE DON'T KNOW WHY

Body paralysis
During REM sleep, muscles are paralysed, so you don't act out your dreams. You can wake in this stage. During this scary experience, you are semiconscious but unable to move.

After a good night's sleep, sleep pressure is low

Adenosine broken down during sleep

Sleep pressure
The longer you stay awake, the larger your sleep pressure. This pressure is comprised of rising levels of chemicals such as adenosine, which causes tiredness by inhibiting neurons in the brain. More adenosine is produced if you've had an active day.

Sleep pressure peaks at bedtime

Unlikely to enter REM sleep upon falling asleep

7AM
6AM
5AM
4AM
3AM
2AM
1AM
12PM

LEVEL 4
LEVEL 3
LEVEL 2
LEVEL 1
REM
AWAKE

DEEP SLEEP
LIGHT SLEEP

AVOIDING SLEEP

Many of us use caffeine to help keep us awake. It makes us more alert by blocking a chemical in the brain called adenosine, which is responsible for making us sleepy. After the effects expire, we suddenly feel very tired.

Range of effects
If you don't sleep you will suffer from a range of physical and cognitive effects. Long-term sleep deprivation can even cause hallucinations.

FORGETFULNESS

LOSS OF RATIONAL THOUGHT

RISK OF ILLNESS

HIGHER HEART RATE

MUSCLE TREMORS

Stages of sleep

Each night you pass through different sleep levels. Level 1 is between sleep and wakefulness. In this stage, you may twitch as muscle activity slows down. As you enter true sleep, Level 2, your heart rate and breathing become even. During deep sleep, Levels 3 and 4, your brain waves slow and become regular. You tend to enter bouts of REM sleep once you have passed through other sleep levels. In REM sleep, your heart rate increases and brain waves look similar to when you are awake.

If you don't sleep

Going without sleep for a long time causes unpleasant symptoms. When you grow tired, your brain will steadily become unresponsive to neurotransmitters (chemicals) involved in regulating happiness. This is why tired people are often moody. When you sleep, your brain resets itself, and becomes sensitive to these neurotransmitters once again. The effects of sleep deprivation become progressively worse the longer you stay awake.

Key
Illustrated here is a typical 8-hour night's sleep. You climb and fall between different levels of sleep in 90-minute bouts, interspersed with REM.

 Awake

 REM sleep

 Level 1 sleep

 Level 2 sleep

 Level 3 sleep

 Level 4 sleep

||||| Sleep pressure

Entering your dreams

Your brain draws on and remixes your memories of people, places, and emotions to create sometimes complex and usually confusing virtual realities known as dreams.

Creating dreams

During REM sleep, your brain is far from asleep. It is highly active in this level of sleep, and it is when you do most of your dreaming. Areas of the brain associated with sensation and emotions are particularly active when you dream. Your heart and breathing rates are high because your brain consumes oxygen at a similar pace to when you are awake. Dreaming is thought to be linked to how the brain processes memories.

Sleepwalking and talking

Sleepwalking occurs during slow-wave, or deep, sleep. At this level of sleep, your muscles are not paralyzed, as they are during REM. The brainstem sends nervous signals to your brain's motor cortex, causing you to act out your dreams. It is more common when people are sleep deprived. Sleep talking occurs during REM sleep if nerve signals that usually paralyze your muscles are interrupted, temporarily allowing you to vocalize in your dreams. It may also happen when you are moving from one level of sleep to another.

Motor area of brain is active

Speech area of brain is active

SLEEPWALKING

SLEEP TALKING

2 HOURS
THE ESTIMATED TOTAL
TIME YOU SPEND
DREAMING EACH NIGHT

NO RATIONAL THOUGHT

Logic impaired
The prefrontal cortex of your brain, where most of your rational thinking occurs, is inactive. You tend to accept crazy events in your dreams as if they are normal, because your dreaming self is unable to process these events as anything else.

NO SENSORY INPUT

Reliving sensations
Your brain receives little new sensory input when you are asleep, so the part of your brain that processes sensory signals is inactive. You do "sense" in your dreams, but you are re-experiencing sensations you had at some point when you were awake.

REM sleep
Nervous signals in the brainstem regulate brain activity during REM sleep. Interactions between "REM-on" and "REM-off" nerves control when and how often you pass into REM sleep. The muscles that move your eyes are the only muscles that are active in REM sleep, so your eyes move when you dream.

RAPID EYE MOVEMENT

BODY PARALYZED

Inability to move
The motor cortex, which controls conscious movement, is inactive. The brainstem sends nervous signals to the spinal cord, initiating muscle paralysis, which prevents you acting out your dreams. The production of neurotransmitters that stimulate motor nerves is completely shut down.

MEMORY CONSOLIDATION

Sleep is important for memory storage. You are more likely to retain new information after you have slept. Dreams are thought to be a by-product of your brain processing and shuffling new memories and forgetting unimportant ones.

Memory forgotten — Shuffled memory

Emotions run wild
The emotional centre in the middle of your brain is highly active, which explains the flurry of emotions you may experience when dreaming. This area encompasses the amygdala, which can be active during nightmares as it regulates your response to fear.

EMOTIVE RESPONSE

SPATIAL AWARENESS

Feeling of movement
Even though you don't move when you dream, you may feel like you do. The cerebellum, which controls your spatial awareness, may become active, resulting in the feeling that you are running or falling in your dreams.

PREFRONTAL CORTEX

MOTOR CORTEX

SENSORY AREA

EMOTIONAL CORTEX

VISUAL CORTEX

CEREBELLUM

BRAINSTEM

MENTAL IMAGERY

Remixed memories
The visual cortex at the rear of your brain is active, because it generates the imagery you experience in your dreams from remembered events. This can include places you've been, people you've met, and even objects you've interacted with. They can either be things you are emotionally attached to, or just completely random.

All emotional

Emotions influence our decisions and occupy much of our waking lives. Social bonds were vital for our ancestors' survival, so we have evolved to be able to read emotions in others. Understanding how emotions work has led us to believe we can influence what we feel.

Basic emotions

A few basic emotions are universally identified. Happiness, sadness, fear, and anger seem to have facial expressions that are recognizable to people in the most widely separated cultures. Combining these gives rise to the huge number of complex emotions we experience.

Fear and anger

The bodily reactions for fear and anger are very similar, even though they involve different hormones. It is mainly your brain's interpretation that determines whether you feel angry or afraid.

Happiness and sadness

Your brain and large intestine produce hormones including serotonin, dopamine, oxytocin, and endorphins, which affect happiness. Lower levels of these hormones result in sadness.

WHY DO WE CRY WHEN WE ARE SAD?

When you are feeling sad or stressed, the tears you shed secrete stress hormones such as cortisol, which is why we feel better after a good cry!

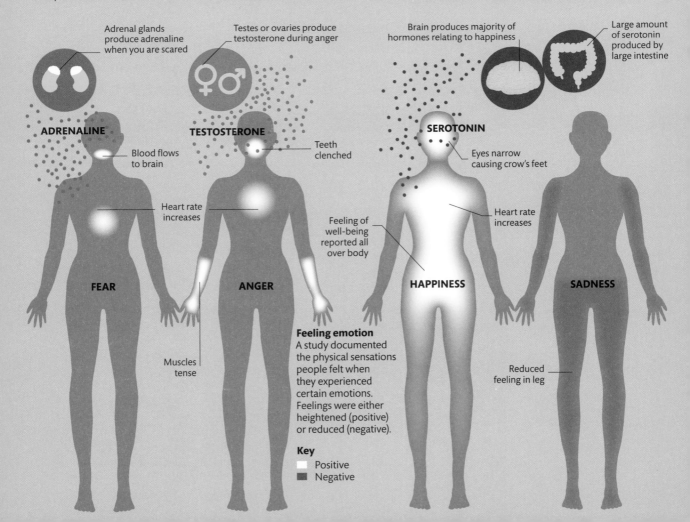

Adrenal glands produce adrenaline when you are scared

Testes or ovaries produce testosterone during anger

Brain produces majority of hormones relating to happiness

Large amount of serotonin produced by large intestine

ADRENALINE

TESTOSTERONE

SEROTONIN

Blood flows to brain

Teeth clenched

Eyes narrow causing crow's feet

Heart rate increases

Heart rate increases

Feeling of well-being reported all over body

FEAR

ANGER

HAPPINESS

SADNESS

Muscles tense

Reduced feeling in leg

Feeling emotion

A study documented the physical sensations people felt when they experienced certain emotions. Feelings were either heightened (positive) or reduced (negative).

Key
- Positive
- Negative

Motor Cortex

Emotional centre of brain

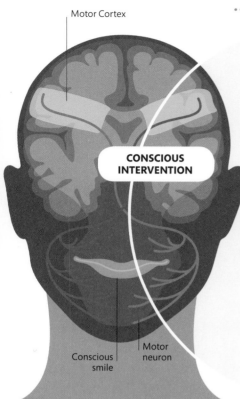

CONSCIOUS INTERVENTION

FEELING

SIGNALS

EXPRESSION

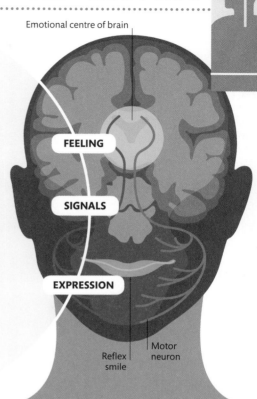

Conscious smile | Motor neuron

Reflex smile | Motor neuron

How emotions form
Emotions consist of feelings, expressions, and bodily symptoms. It may seem like your feelings come first, however a feedback loop allows the body to regulate your emotions and vice versa. At a certain point in this cycle you are able to reinforce, inhibit, or change emotions by altering your response. For example, if you are feeling happy, continuing to smile will make you feel even happier!

Conscious facial expressions
After you have started to experience an emotion, you are able to change your facial expression to hide or reinforce your true emotion. This action is consciously controlled by neural pathways from the motor cortex.

Reflex facial expressions
When you experience emotion, facial expressions appear without your control. For instance, when you hear good news, you cannot help but smile. These reflex actions are thought to be due to signals from the amygdala, in the brain's emotional centre.

THE HAPPINESS YOU FEEL DURING A "RUNNER'S HIGH" IS CAUSED BY NATURAL CHEMICALS IN THE BRAIN CALLED OPIOIDS.

WHY DO WE HAVE EMOTIONS?
Experts think that emotions evolved as a pre-verbal way of communicating. By understanding emotional signals, we can form stronger social bonds. Facial expressions can demonstrate you are in need of help, sorry for something you have done, or can warn others to stay away if you are angry. However, some scientists think there is a simpler explanation; the widened eyes of fear could help us to see better, and the wrinkling of the nose in an expression of disgust could be a way of rejecting harmful chemicals in the air.

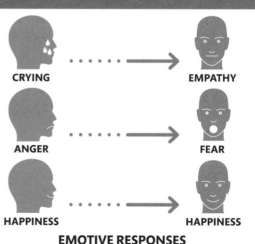

CRYING → EMPATHY

ANGER → FEAR

HAPPINESS → HAPPINESS

EMOTIVE RESPONSES

Fight or flight

When we are threatened, our body springs into action. Our brain sends signals to the body causing a variety of physiological changes that prepare us to face the challenge or to run away.

Activating a response

Have you ever been startled by a garden hose, only to realise it is not a snake and completely harmless? Before we are even consciously aware of a threat, our brain activates the nervous system, which causes the release of hormones from the adrenal glands. Meanwhile, the information also travels the longer route to our cortex where conscious brain regions can analyze whether the threat is genuine. If not, it will calm down the physical reaction.

SNAKE

IN TIMES OF **HIGH STRESS** YOU MAY EXPERIENCE **TUNNEL VISION,** IN WHICH YOU **DON'T NOTICE WHAT HAPPENS AROUND YOU**

CORTEX

VISUAL CORTEX

Visual cortex processes the image after automatic reaction

THALAMUS

Thalamus passes sensory information as nervous signals to amygdala

HIPPOCAMPUS

AMYGDALA

Amygdala activates nervous response and instructs pituitary gland to release hormones

Pituitary gland releases adrenaline and cortisol

1 Brain activity

The amygdala signals the body to take action before the fearful stimulus has even been recognized by the visual cortex – this happens when you jump at shadows. Then, the visual cortex fully analyzes the image to check if the threat is real, and your physical reactions adjust accordingly. Your cortex also consults memories stored in the hippocampus to check if the threat was faced in the past.

ADRENALINE AND CORTISOL

NERVOUS SIGNAL

2 Alternative pathways

Signals from the brain are sent to the body via nerves, and also by hormones released from the pituitary gland. The nervous signals travel faster than the hormones, so they kickstart hormone production in the adrenal glands.

IMMUNE SYSTEM ACTIVITY REDUCED

FAT USED FOR ENERGY

HIGH BLOOD SUGAR

5 Long-term effects
Over minutes and hours, signals from the adrenal glands continue to cause a cascade of reactions. Blood sugar rises and fat stores are metabolised for energy so your muscles continue to work at their full potential. Non-vital processes, such as immune system activities, are shut down to conserve energy.

ADRENAL GLANDS

MODERN STRESS

Modern stress tends to be very different to the type encountered by our ancestors – our stressors often overstay their welcome, and can't be dealt with by fighting or fleeing. Stress is helpful in the short-term, but continued stress negatively affects your health, causing headaches and illness.

PERSISTENT STRESS

IMMEDIATE STRESS

3 Hormone producer
The adrenal glands that sit on top of the kidneys produce more adrenaline and cortisol in response to the nervous signals and hormones sent by the pituitary gland. This heightens the physical effects of stress.

BLOOD VESSELS CONSTRICT

BLOOD FLOWS TO MUSCLES

HEART RATE INCREASES

BREATHING RATE INCREASES

PUPILS DILATE

4 Short-term effects
Within seconds, heart rate and breathing increase to boost oxygen circulation. Blood vessels close to the skin constrict, leaving you pale, and your bladder muscles relax, possibly leading to embarrassing accidents!

Emotional problems

Our emotions are controlled by a balance of chemicals and circuitry in the brain, so imbalances of certain chemicals can cause emotional disorders. Experts once believed they were purely psychological, but they now understand that physical changes underlie each illness.

Phobias

A fear is classed as a phobia if the fear outweighs the threat. It is logical to be wary of snakes, which can be deadly. If that fear extends to pictures or toys and begins to affect daily life, it becomes a phobia. Phobias can develop over time, be learnt at an early age, or be associated with an incident involving the stimulus.

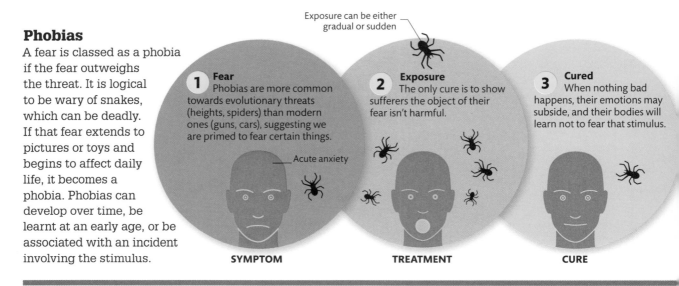

Exposure can be either gradual or sudden

1 Fear
Phobias are more common towards evolutionary threats (heights, spiders) than modern ones (guns, cars), suggesting we are primed to fear certain things.

Acute anxiety

2 Exposure
The only cure is to show sufferers the object of their fear isn't harmful.

3 Cured
When nothing bad happens, their emotions may subside, and their bodies will learn not to fear that stimulus.

SYMPTOM **TREATMENT** **CURE**

Obsessive compulsive disorder

Sufferers of obsessive compulsive disorder (OCD) experience intrusive negative thoughts leading to compulsive behaviours, which they incorrectly believe can relieve their anxiety. OCD is possibly caused by overactivity in the areas connecting the brain's frontal lobe to deeper areas. Most cases are manageable with treatment.

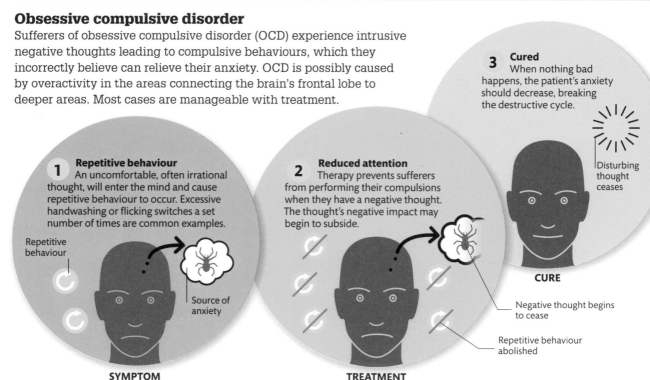

3 Cured
When nothing bad happens, the patient's anxiety should decrease, breaking the destructive cycle.

Disturbing thought ceases

1 Repetitive behaviour
An uncomfortable, often irrational thought, will enter the mind and cause repetitive behaviour to occur. Excessive handwashing or flicking switches a set number of times are common examples.

Repetitive behaviour

Source of anxiety

2 Reduced attention
Therapy prevents sufferers from performing their compulsions when they have a negative thought. The thought's negative impact may begin to subside.

Negative thought begins to cease

Repetitive behaviour abolished

CURE

SYMPTOM **TREATMENT**

TRAUMATIC MEMORIES

After trauma, some people experience flashbacks, hypervigilance, anxiety, and depression – these are the symptoms of post-traumatic stress disorder (PTSD). When you are afflicted, recalling the traumatic memory will trigger a "fight or flight" response, unlike ordinary memories. Treatments can be provided through therapy or drugs.

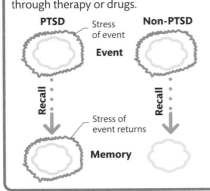

PTSD   Stress of event   **Non-PTSD**

Event

Recall   Recall

Stress of event returns

Memory

BRAIN ACTIVITY

Thalamus active due to linking once pleasant stimuli with negative emotions

Emotional centre of the brain is highly active, dealing with anger, sadness, and pain

Activity in the prefrontal cortex reduced, affecting concentration, memory, and processing

Depression

The symptoms of depression include low mood, apathy, sleeping problems, and headaches. It is thought to be caused by chemical imbalances in the brain, leading to certain areas becoming overactive or underactive. Antidepressants can help re-set this balance by raising levels of chemicals, but they only address the symptoms, not the cause. Attitudes towards depression have progressed to understanding it as a condition, not a state of mind.

Bipolar disorder

Featuring changes in mood from mania to extreme depression, bipolar disorder is highly genetic – it runs in families – but it is often triggered by a stressful life event. Bipolar disorder is a sub-type of depression. It is thought to be due to problems with the balance of certain chemicals in the brain, including noradrenalin and serotonin, and this causes the brain's synapses to become either overactive in mania or underactive in depression.

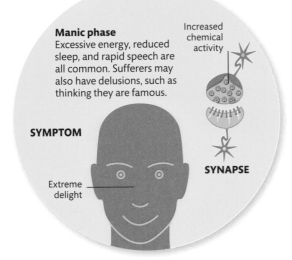

Manic phase
Excessive energy, reduced sleep, and rapid speech are all common. Sufferers may also have delusions, such as thinking they are famous.

Increased chemical activity

SYMPTOM

Extreme delight

SYNAPSE

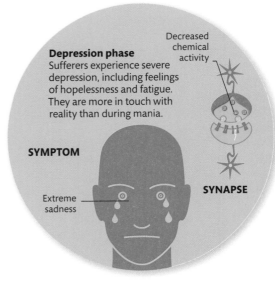

Depression phase
Sufferers experience severe depression, including feelings of hopelessness and fatigue. They are more in touch with reality than during mania.

Decreased chemical activity

SYMPTOM

Extreme sadness

SYNAPSE

Feeling attraction

Scientists are only just beginning to understand what happens when we feel attracted to someone, why we are attracted to certain people and not others, and why we make our choices – and it is mostly to down to hormones.

Chemical bond

When attraction begins, hormones play an important part in augmenting our romantic feelings. Levels of dopamine in the brain increase, providing the familiar rush of pleasure. A chemical that is converted into adrenaline is released, causing a dry mouth and sweaty palms. It also causes your pupils to widen, which signals your desire to the other person, making you increasingly attractive. Serotonin levels change and are believed to lead to obsessive, lustful thoughts.

DOES CULTURE AFFECT ATTRACTION?

Within a single culture, ideals of beauty change over time. In Europe, pale skin and a full figure once indicated wealth and was typically seen as attractive in a woman. Now, a thinner, more tanned figure is seen as desirable.

Arousal-initiation area

Ventromedial prefrontal cortex

Dilated pupil

Heart rate increases as attraction grows, so we may confuse the sensations of love and fear, making a scary film a great first date!

FACIAL SYMMETRY

SENSE OF HUMOUR

BODY SHAPE

TONE AND SPEED OF VOICE

COLOUR OF CLOTHING

1 Immediate lust
Within moments of seeing someone you are attracted to, an area of the brain called the ventro-medial prefrontal cortex is activated to analyze their dating potential. Testosterone is released in both genders, stimulating feelings of lust.

2 Contributing factors
Our attraction uses cues such as facial symmetry and body shape, as they signal good health and fertility. Other cues, such as similar interests, highlight whether we are compatible in the long term. The colour red ignites passion in both sexes.

PROLONGED EYE CONTACT **INCREASES THE MAGNETISM** BETWEEN TWO PEOPLE

3 Long-term pair bonding

After the initial attraction phase, relationships change, and a different set of hormones become important. Oxytocin is released after sex, and increases feelings of trust and bonding, which aids in establishing relationships. Another hormone, vasopressin, is equally important. It is released when two people spend a great deal of time together, promoting monogamy.

SEX

BODY ODOUR

Sweat can tell us how healthy someone is, and even whether we are genetically compatible. People who have an immune system relatively different to our own tend to smell more attractive, since a mixing of these genes would encourage healthier offspring. Generally, women prefer the scent of men somewhat similar to their own over the scent of those genetically identical or completely dissimilar.

OVULATION

Changing signals
When women are ovulating, there are subtle changes that indicate fertility; voice pitch rises, cheeks flush, and you tend to flirt more and dress more attractively.

MENSTRUAL CYCLE

Subtle signals

In many animals, it is obvious when females are fertile, through bold signals such as brightly coloured swellings on their body or pheromones in their urine. When it comes to humans, ovulation isn't so obvious – and it's not known why we evolved that way. Nevertheless, women do have subtle ways of advertising their fertility, such as flirting more and dressing more attractively, and men seem subconsciously to be able to pick up on these signals. One study showed that men release more testosterone when reacting to the scent of women who are ovulating than those in a less fertile phase of their cycle.

Extraordinary minds

Everyone's brain is unique, but there are some people who can do amazing things that most of us can only dream of. Slight changes in the wiring of the brain, or the way we learn to use it, can give rise to these incredible abilities.

Delayed language
Children with autism (but not Asperger's) take longer to learn language, and some never become verbal. Those who do speak may have trouble using words to communicate with others as an adult.

Socialising impaired
Reduced eye contact is an early sign of autism. Autistic individuals tend to dislike socializing, finding its complex rules confusing and frightening. Nevertheless, this is not to say those with autism never form strong social bonds.

Repetitive behaviour
People with autism process information differently, and this means everyday situations can be overwhelming. Self-soothing, routine behaviours are common, and can help people with autism calm themselves when anxious.

Specific interests
Those who are autistic often develop narrow, specific interests. These can be a source of comfort and enjoyment, possibly because the structure and order of familiar topics provides respite from the confusing social world.

SOMETIMES AUTISM LEADS TO

Autism spectrum

Autism spectrum disorders (including Asperger's syndrome) are probably caused by unusual patterns of connectivity in the brain. Genes are known to play a role as autism runs in the family, although why they affect some people only mildly while others need care throughout their lives isn't known.

Rare prodigious qualities
Occasionally, those with autism show incredible skills in areas such as maths, music, or art. This may be due to a characteristic pattern of brain processing that focuses on details.

Increased connections
When any brain grows, non-essential nerve cell connections are pruned. It is thought that in autism, this process is inhibited, resulting in too many connections.

SENSORY SHORT CIRCUITS

Some people have cross-overs between their senses. Some see letters or numbers as coloured while others might taste coffee when hearing a C-sharp. Their condition is called synaesthesia and it happens because they do not undergo the same nerve-cell pruning process that other people do during their childhood brain development. The result is extra connections between the brain's sensory areas. Synaethesia is thought to be genetic as it tends to run in families. However, since some identical twins have synaesthesia while the other twin does not, genetics cannot be the whole story.

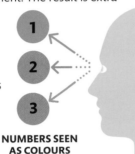

NUMBERS SEEN AS COLOURS

Hallucinations

Hallucinations are surprisingly common; many recently bereaved people report seeing their spouse, and almost everyone has seen something non-existent out of the corner of their eye. These are a normal by-product of our brains' attempts to make sense of the world.

EXPERIENCING HALLUCINATIONS

Types of hallucinations
You may think somebody called your name, but no such thing was said, or you may see a shadow out of the corner of your eye. These are all common types of hallucinations.

BY THE **AGE OF 5,** THOSE WITH **SUPERIOR AUTOBIOGRAPHICAL MEMORIES** START TO **REMEMBER EVERYTHING**

Memory champions

Some people have amazing memories, but they mostly use techniques such as placing the items that need remembering along a familiar route. A handful of people with a condition called superior autobiographical memory automatically remember every insignificant event that has happened to them for their entire lives. One individual with this condition had an enlarged temporal lobe and caudate nucleus – both areas of the brain that are linked to memory.

NEW NEURAL CONNECTIONS

Memory pathway
If you need to remember a sequence of numbers, one way to do so is to associate each number with a place or object you see on your journey to work. Fitting a "3" in the window of a car or building, for instance, helps to retain that number in place in the sequence.

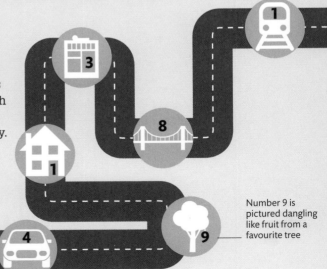

Number 9 is pictured dangling like fruit from a favourite tree

Index

Acknowledgments

DK would like to thank the following people for help in preparing this book: Amy Child, Jon Durbin, Phil Gamble, Alex Lloyd, and Katherine Raj for design assistance, Nadine King, Dragana Puvacic, and Gillian Reid for pre-production assistance, Caroline Jones for indexing, and Angeles Gavira Guerrero for proofreading.

The publisher would like to thank the following for their kind permission to reproduce their images:
p.85: Edward H Adelson
p.87: Photolibrary: Steve Allen

For further information see:
www.dkimages.com